# 118 Theories of Design[ing]

Edited by

**Paul A. Rodgers**
University of Strathclyde, UK

**Craig Bremner**
Charles Sturt University, Australia

**Series in Art**

VERNON PRESS

www.vernonpress.com

In the Americas:
Vernon Press
1000 N West Street, Suite 1200
Wilmington, Delaware, 19801
United States

In the rest of the world:
Vernon Press
C/Sancti Espiritu 17,
Malaga, 29006
Spain

Series in Art
Library of Congress Control Number: 2020947495
ISBN: 978-1-62273-962-2

# Contents
# 118 Theories

# About the Authors

Paul Rodgers is Professor of Design in the Department of Design, Manufacturing and Engineering Management (DMEM) at the University of Strathclyde, UK. He is also the Arts and Humanities Research Council Leadership Fellow for Design in the UK (2017 - 2020). He is a co-founder of the Design Disruption Group, with Giovanni Innella, Freddie Yauner and Andy Tennant, who strive for positive change in health and social care and elsewhere.

Craig Bremner is Professor of Design at Charles Sturt University, Australia. His research deals with developing methods to discover and to value why 'not-knowing' is an essential beginning point of design practice.

Paul is from Glasgow, Scotland, and Craig is from Sydney, Australia. We met at Northumbria University where, on the same day in November 2009, we started working for the School of Design. In the new, large open-plan office in the new, large Design building, we were allocated the last two vacant seats, side by side, in the farthest corner of the office. Even though Craig had worked in Glasgow for several years and Paul had contemplated a very good job offer in Sydney, we hadn't met until this moment. And a bit like schoolboys seated together randomly in a very traditional Design School, we quickly discovered we had a very well informed, but mutinous, take on Design.

With that shared knowledge and attitude, we began our collaboration with a paper for a conference on the PhD in Design, where we urged the candidate to be undisciplined and irresponsible. Soon after, we wrote our first Charter - "Design School: Undisciplined and Irresponsible" - which we have performed several times at conferences, often by invitation. To us, Design is now so affected that sometimes we cannot resist writing a response to calls that simply invite mutiny. More often we send each other ideas or challenges to fracture the commonplace Design promises. When an idea takes hold, we write and then search for a publisher. We get a bit peeved when what we write is rejected by orthodox reviewers who want affirmation, not critique. Despite this reaction, everything we write finds its way into publication. We also write in relay from different countries and time zones - one writes until we have had enough and sends it to the other and so on - which means that often one works while the other one sleeps. If the morning email has an attachment, then we have to do something. If not, we take it easy. Working together this way appears to us to be inexhaustible and has produced a lot of publications over the years (see the list below).

More recently, we have embarked on a series of international workshops that focus on the relationship between the act of Design and the gesture of Care. Something that is very dear to us. At the first of these workshops, we produced our second Charter - "The Lancaster Care Charter" - a format we like. Our main challenge is to always remain challenging.

**Books**

Design School: After Boundaries and Disciplines
Design School: The Future of the Project
Does Design Care…?! Head-to-Head Debates
An Illustrated A to Z for the Design of Care
Does Design Care…? An International Workshop of Design Thought and Action

**Charters**

The Lancaster Care Charter
Design School: Undisciplined and Irresponsible

**Book Chapters**

The Design of Nothing: A Working Philosophy

**Journal Papers**

A is for Anthropocene: An A–Z of Design Ecology
Paradoxes in Design Thinking
The Concept of the Design Discipline
An A to Z of Contemporary Design
Exhausting Discipline: Undisciplined and Irresponsible Design
Design Without Discipline
Alterplinarity - "Alternative Disciplinarity

**Conference Papers**

Design School: Undisciplined and Irresponsible

Design School: Design Education in the Age of Digital Capital

Alterplinarity: The Undisciplined Doctorate and the Irresponsible Candidate

# Foreword

Rachel Cooper

Paul Rodgers and Craig Bremner will explain that their collaboration and writing in design is led by rejecting orthodoxy. They write together to present "*ideas or challenges to fracture the commonplace that Design routinely promises*". So, one might ask why a book of 118 Theories of Design(ing)? As both a practising designer and now a design academic, I have always held a certain antipathy towards theory and theoreticians. And, to some extent, like the author and commentator Will Self (2019) have been "Against Theory", because in his words theories (especially in the Arts, Humanities and Social Sciences ) are "*...fences of impenetrable prose designed to keep people out...*" Actually, building on what Max Weber (1978) described as social closure and the tendency of groups to restrict entry to outsiders by building barriers, Design is not (should not) be a profession or a discipline built on that premise.

This is not a book designed to keep people out, nor one which aims to build a grand theory upon which we can predict the future, such as Charles Darwin's Theory of Evolution by Natural Selection or Albert Einstein's Theory of General Relativity. The basis for this work is that "*...all design practice (designing) is embedded in discourse*" and thus, these 118 Theories of Design(ing) are, in effect, a proposition, a challenge to orthodoxy. The first challenge is that the contributions are not long evidence-based arguments set up in an abstract language theory. They are short, philosophical, sometimes flippant, argumentative and challenging statements. 118 Theories of Design(ing) is written not only by Paul Rodgers and Craig Bremner, but by many contemporaries working in design that they have happened upon during this compilation journey.

The second challenge is that, unlike science, where theories are created to explain the past or the present and predict the future, the approach to theories of design(ing) taken here, is one that helps us to contest the future, and our fixation on designing the possible and to consider how we might give form to the impossible. Here as Paul Rodgers and Craig Bremner state: "*...theory becomes the de facto critique of practice...*", and therefore theory here is the beginning and not the end.

On this basis, if we consider one of the entries - #92 A Scientific Theory of Design[ing] in which Peter Lloyd Jones states that "*...if design exists in the future, then neither the current spaces used to focus it nor the names used to frame it currently exist either. By way of extension, if we accept that design is a nameless, non-discipline that exists in the future within an imagined, but currently unknowable space, where it is made real, then it is plausible to speculate upon a theory of design that is boundless and pre-disciplinary.*" Perhaps so and indeed we are presented with the beginning not the end, but with 118 Theories of Design(ing) to be used to spread the imaginary of design(ing) and enable the boundless, unrelenting combination of theory and practice to work with confidence in the unknowable space that is the future.

As I write this foreword, we are in week four of the COVID-19 lockdown in the UK. Never in my lifetime has there been such a drastic social and economic event. Most designers working today have not experienced an event that has such local and global impact; community connection, but increased poverty, lower energy emissions, improved climate but smaller economies. We are undergoing a global reset, corporate and individual reflection on how to survive, what is survival and what it means to be human. Tomorrow, next month, next year are unknowns. For many, we hope we don't go back to the 'past normal', but we find a 'new normal' or an 'improved normal' for the sake of the planet and it's inhabitants. Reflection, observation, discourse, critique, challenging everything and taking action is still the baseline for design(ing). Let's not use the theories here "*...as a hose to water the scholarly allotment...*" (Self, 2019) but to design the impossible and the unthinkable to create a better world post-Coronavirus.

## References

Will Self, *A Point of View: Against Theory*, BBC podcast 16 August 2019.

Max Weber, *Economy and Society: An Outline of Interpretive Sociology*, Edited by Guenther Roth and Claus Wittich, Berkeley: University of California Press, 1978.

# Introduction

Paul Rodgers & Craig Bremner

*"The time of theory is today."* Slavoj Žižek

If, according to the French poet Paul Valéry, a philosophy should be portable, then through this book, we illustrate how a theory should be palpable - perceptible to the senses but impractical to systemization. However, we did apply a system to its composition. Each theory is around 140 words (more or less) and that was a recoil from our initial challenge of 140 characters - the initial length of a tweet; once a benign and sweet description of birdsong but now so toxic we felt it was unseemly. We might give the impression we have also exchanged chirping for carping, but so little of what is 'in theory' possible is made actual by design that clearly, something is missing. This book feels the loss of the dream of design and compensates by making a contribution to the sizable account of what is missing.

Foucault has an interesting notion about the writing of books. He says "*I don't write a book so that it will be the final word; I write a book so that other books are possible, not necessarily written by me.*" (O'Farrell, 2005: 9). Paraphrasing Foucault, we might be tempted to claim that the aim of this book is to once again make possible the future scenarios so essential for the existence of design(ing). But we have also written this book to make clear we are not concerned with the future nor the scenario, but the possibility of the possible – that design is possible. This book isn't entirely written by us. We have assembled it to also make possible an obsolete meaning of theory; to illustrate a mental view of design(ing); to expand the 'mental space' or more fashionably to spread the 'imaginary' of design(ing).

## The concepts, insights and arguments behind 118 Theories of Design[ing]

Why do we need theory? Where do theories come from? Is the parable "*...design theorists [as opposed to design practitioners] are those that cannot design*" justified? Are the only design theories worthy of consideration those that are rooted in practice? What can one hope to get from design theory? These questions and others are laid out in Gui Bonsiepe's wonderful book

"Interface – An Approach to Design". We cannot, of course, expect there to be a single clear answer to these questions.

Bonsiepe, however, articulates one solid argument in favour of design theory. That is, that all design practice (designing) is embedded in discourse. For Bonsiepe, theory and practice have always been closely interlinked. Moreover, design discourse exists amongst universes of linguistic variations that are all part of an indispensable part of practice. A world of design that focuses solely on practice and neglects theory endangers the act of designing to the shadows, "*...which sharply contradicts the cultural and economic importance of design as a central domain of Modernity.*" (Bonsiepe, 1999: 22)

We contend that design theory is important to the global culture of design but given its relative disregard, the concept of this book is to give theory a chance. Like the histories of design, design(ing) doesn't seem to see much future in the past, and theories (or speculations) don't seem to have much currency even while design(ing) seems continually intoxicated with possible futures. In this book we show how design(ing) is constantly rubbing against theories (and vice versa) because for every possible future there is a future impossible – for the irresistible promise of what-might-become there is increasingly the paradox of what-might-not-become.

When Terry Eagleton (2018) states in typically enigmatic style "*...the only image of the future is the failure of the present*" he exposes an inconsistency in design. A design fundamental has always been its ability to discern between the possible (what-might-become) and the actual (what-might-not-become). Design is nourished on the belief that almost anything is possible. But the infinite scope of the possible is limited only by the materialisation of the actual. Because of its addiction to the infinitely possible, the problem for design is almost anything is possible except the actual. This problem exposes a tension between theory and practice that seems to lie in wanting to comprehend a paradox that has vexed both (but mostly practice); namely, resolving the difference between what design does and what design only dreams of doing. This paradox can only find resolution in texts that are mostly disparagingly labelled theory.

The difference between a desirable and an actual state of affairs adds another dimension to theory; that of critique where the tension between theory and practice takes on a new complexion. In the absence of a platform of critique of design, theory becomes the *de facto* critique of practice. Under the banner of critique theory thus becomes a beginning, not an end. In this scenario theory functions so you can read the operating instructions.

If design is about a possible future (or future possible) does theory come before or after all the attempts by design at putting this possibility into practice – i.e. is theory a history or a speculation? By definition theory is a speculation that habit tells us is derived from contemplating the world-as-found. But Flusser (1988: 17) maintains that "*...theory may be understood not as the contemplation of form but of the shaping of it.*" In that sense, design theory is not knowledge gained from the act of designing what-might-become. Neither is theory a service to practice providing insights into designing. Following Flusser's train of thought and contrary to the habitual use of theory as a way of explaining why the world looks like it looks (Foucault's "what do does"), theory is a projection giving shape to the world. Flusser's astute inversion of the role of theory is of critical importance for designing because having exceeded the carry capacity of the world 'what-might-become' has become 'what-might-not-become'. Design is having to give form to the impossible – which for design is paradoxical (Rodgers, Innella and Bremner, 2017).

## Origins

The word theory has its roots in the late 16th-century Greek term *theoria* meaning 'contemplation' or 'speculation' – two words that frequently feature in contemporary design discourse. Theory, the dictionary tells us, is "*a supposition or a system of ideas intended to explain something*" (e.g. Darwin's theory of evolution). In this book of 118 theories of design[ing], however, we do not seek to explain anything. Rather we ask the reader to read the 118 theories of design[ing], reflect upon them and question everything...
We chose to write our theories using a limited word count in order to solicit a large number of theories, all of which give the reader ample room to explore the ideas, source their genesis, and expand them as seems to fit the thoughts and actions of design(ing).

## Its engagement with or implications for practice and thought

The basis for a theory is the spectacle, meaning both a public show and to look, but if we accept the common notion of theory is 'to speculate' we also have to accept that 'to speculate' is also to play with financial risks for profit. There is then one implication for theories of design(ing) – dealing with the results of practice risking the carrying capacity of the planet with unnecessary stuff and unfulfilling services while chasing profit. In this book, we do not take a position on speculation but we position speculation as an everyday element in all design thought and action.

## A sense of the substantive or theoretical issues that might be engaged

The belief that design comes from somewhere is closely linked with the belief that design is going somewhere (e.g. what design can do). Our view of design in this book reflects the various authors' views of society. And as everyone now uses the same interfaces resulting in the blurring of disciplines, the theories in this book aim to re-focus our sights on the lost subject of design(ing). Each theory can be read individually and/or randomly, and now all theories have further reading resources that we have listed. Read as a whole the substantive contribution of the book is a restored picture of why design(ing) looks like it looks. Having surveyed this restored picture, it is then impossible for the reader to avoid asking what needs to be done? And we imagine this question will be answered differently by every reader.

## How concrete examples might be drawn into the argument

Each theory is a concrete example of theory-in-action as the essential 'carnival' mirror for practice continuously distorting what-is to reflect new ideas about what-might-or-might-not-become.

## How proposals for action might be derived from the arguments

Design is where theory and practice meet. The separation of theory and practice is another false dualism that we have to learn to overcome. By classifying initiatives as either theoretical or practical, we are not paying attention to the fact that our view of the world is already deeply informed by theories about the world. In saying we don't have time to waste with theoretical considerations, let's get practical and start implementing solutions, what we are actually implying is that there is no need to question our perspective and explore alternative perspectives. We are jumping straight into action, offering answers to the questions and solutions to the problems at hand, without stepping back to make sure we are asking the right questions. We fail to explore whether the solutions we are aiming for are yet again solving one issue whilst causing harm and ugliness elsewhere.

Every practical act is deeply informed by a whole set of theories and perspectives. So, the question is not whether we are practical or theoretical, but rather whether we are implementing practice in full awareness of the theoretical frameworks — the worldview and value systems — that inform our practice. Taking a design-based approach can help us to make our practice more theoretical and our theory more practical.

Design[ing] involves activities and processes that lie at the intersection of theory and practice. Design[ing] is where art and science meet. Design[ing] integrates and exploits information from many disciplines where we have separated human knowing and doing. Design[ing] is where we can acknowledge the influences of the past and give rise to visions of different futures.

The book provides a conceptual basis of the subject and area of study – design(ing) – and as such is a classic text on theory. Even more 'classic' is our approach, which is intended to challenge or provide an alternative to critical methods and interpretations that are established and traditional. But we have reversed the 'classic' origins of theory by taking the practical experience and activity of design(ing) and turned it into explanations of somewhat overlooked and undervalued essays, papers, book articles, words, terms, authors and phenomena that swirl around design(ing), most of which are commonplace and therefore now functions of the mercantile spectacular.

## References

Gui Bonsiepe, *Interface – An Approach to Design*, Maastricht: Jan van Eyck Akademie, 1999.

Terry Eagleton, "Seven Types of Atheism by John Gray Review – Is Every Atheist an Inverted Believer?", *The Guardian*, Wed 11 Apr 2018, https://www.theguardian.com/books/2018/apr/11/seven-types-atheism-john-gray-review-atheist-believer-material-world

Vilém Flusser, "On Discovery", *Artforum*, 27, No. 10, Summer 1988: 17.
Clare O'Farrell, *Michel Foucault*, London, UK: Sage Publications, 2005.
Paul A. Rodgers, Giovanni Innella and Craig Bremner, "Paradoxes in Design Thinking", *The Design Journal*, Supplement 1, Volume 20, 2017, pp. S4444-S4458, ISBN 978-1-138-09023-1.
Slavoj Žižek tells Owen Jones: 'Clinton is the problem, not Trump' https://www.theguardian.com/books/video/2018/oct/24/slavoj-zizek-tells-owen-jones-clinton-is-the-problem-not-trump

| #03<br>**Af**<br>Affordable | #08<br>**Ch**<br>Cheap | #10<br>**Hu**<br>Humble | #21<br>**Re**<br>Repairable | #37<br>**Sw**<br>Swatchification | #68<br>**Ju**<br>Jugaad | #69<br>**Ga**<br>Gandhian |
|---|---|---|---|---|---|---|
| #49<br>**Po**<br>Pornographic | #96<br>**Ex**<br>Excessive | #100<br>**Ca**<br>Careful | | #20<br>**Au**<br>Automated | #22<br>**Bu**<br>Business | #46<br>**Fa**<br>FabLab |
| #98<br>**De**<br>Design Thinking | #107<br>**De**<br>Designer-as-Mythologist | | #12<br>**Co**<br>Cosmological | #15<br>**Ru**<br>Rural | #25<br>**Cl**<br>Climatic | #62<br>**Ev**<br>Evolutionary |
| #117<br>**Ge**<br>Gendered | | #16<br>**Ro**<br>Romantic | #32<br>**Je**<br>Jealous | #40<br>**Se**<br>Sentimental | #67<br>**An**<br>Anecdotal | #74<br>**Pr**<br>Proud |
| #57<br>**Ma**<br>Mature | #58<br>**Po**<br>Political | #60<br>**Po**<br>Power | | #18<br>**Tr**<br>Transitive | #55<br>**Da**<br>Data | #56<br>**Ju**<br>Justified |
| | #04<br>**Po**<br>Possible | #07<br>**Fu**<br>Future | #13<br>**Ul**<br>Ultimate | #14<br>**Ut**<br>Utopian | #35<br>**Wi**<br>Winning | #41<br>**Fa**<br>Fantastic |
| #118<br>**Be**<br>Better | | #23<br>**Re**<br>Responsive | #24<br>**Sh**<br>Sharing | #27<br>**An**<br>Anonymous | #52<br>**Th**<br>Theoretical | #54<br>**Ag**<br>Against |
| #106<br>**Pl**<br>Plurality | #112<br>**Pl**<br>Pluriversal | #116<br>**De**<br>Declassified | | #02<br>**Du**<br>Duplicate | #05<br>**Di**<br>Disruptive | #06<br>**Un**<br>Undisciplined |
| #39<br>**Fa**<br>False | #42<br>**Pl**<br>Plastic | #73<br>**Bi**<br>Bipolar | #94<br>**Di**<br>Disempowering | #108<br>**On**<br>Ontological | #109<br>**Ma**<br>Materials' History | |
| #72<br>**An**<br>Anti-Design Thinking | #79<br>**Ir**<br>Irrelevant | | | | | |

| | | | | | | |
|---|---|---|---|---|---|---|
| #70 Ra Rasa | #99 Me Messy | #113 Po Post-Economic | | #01 Lu Luxurious | #28 Co Comfortable | #48 Ex Exaggerated |
| #51 Le Learning | #53 Pr Practical | #61 Ed Educational | #75 St Studio-Based | #77 In Inflation of Design | #87 Co Constructivist | #91 Ea Early Learning |
| #66 An Anthropocentric | #93 Ca Catastrophic | #104 1.5 1.5°C | #105 Cu Cultural etc... | #110 Co Communal | #114 Of Of the South | #115 In Indigenous |
| | #30 Ma Matriarchal | #36 Zo Zombie | #43 Gr Green | #44 Da Dark | #45 Gr Grey | #47 Ba Barbaric |
| #59 Ci Circumstantial | #80 Ev Everything and Nothing | #83 Tu Tuning | #90 No Non-Cartesian | #92 Sc Scientific | #95 Fi Fictional | #97 Ma Management |
| #78 Id Idealist | #85 Sp Speculative | #86 Gl Global Health | #88 Ca Caring | #101 Ag Agile | #103 Sm SMART | #111 Au Autonomous |
| #71 Pl Pluralistically Confusing | #76 In Inclusive | #81 Im Immersive | #82 Fa Failing Forward | #84 Lu Ludic | #89 Mo More than Human(ist) | #102 Lu Ludic |
| #09 Im Imperfect | #11 De Derived | #26 Mi Misfit | #29 Pa Paradoxical | #31 Da Dangerous | #33 Mo Mongrel | #34 Ti Tired |
| #17 Ra Radical | #19 Un Unknown | #38 An Anti-Design | #50 Cu Curious | #63 Un Uncertain | #64 Pe Pessimistic | #65 Th Thoughtless |

# 118 Theories of Design[ing]

# 118 Theories of Design[ing]

Paul Rodgers & Craig Bremner

There are 118 chemical elements in the periodic table. The periodic table, also known as the periodic table of elements, is a tabular display of the chemical elements, which are arranged by atomic number, electron configuration, and recurring chemical properties. In total, 118 elements have been identified. In the periodic table of the chemical elements, the first 94 occur naturally on Earth, and the remaining 24 are synthetic elements. Iron is the most abundant element (by mass) making up Earth, while oxygen is the most common element in the Earth's crust. Chemical elements constitute all of the ordinary matter of the universe, which makes up only about 15% of the matter in the universe. The remainder is dark matter; the composition of this is unknown, but it is not composed of chemical elements

So, 118 elements make up all of the ordinary matter of the universe. The 118 theories of design[ing] in this book are the current known theories that take up all the ordinary observable matter of design. The rest is just dark matter...

The following 118 theories of design[ing] are assembled into 11 thematic groups.

## Meagring

***[verb]***

*something lacking in quantity or quality (of a designed product, service, or experience); lean, agile.*

The meagring theories of designing listed here all connect where something appears lacking in quantity or quality (e.g. of a designed product, service, or experience) or where the theory suggests less of something than there could or should be. Meagring theories suggest that there are not enough (resources) or that what there is may be of poor quality. The meagring theories here also allude to 'thinly dispersed' phenomena where things are in short supply. Meagring theories also refer to things that are inadequate. For example, **#03 An Affordable Theory** warns us against furthering a world that is becoming increasingly unaffordable to anyone who doesn't already own it. **#08 A Cheap Theory** of Design[ing] reminds us that design has always been dying to make things cheap whereas **#10 A Humble Theory** suggests that acknowledging that we might know very little about many things is the first step towards greater enlightenment for design as this will require us to embrace unfamiliarity, inexperience, cluelessness, and an honest lack of knowledge that will be totally refreshing. **#21 A Repairable Theory** of Design[ing] balances the constant technological push for novelty by reminding us to recognise the fragility of design. A positive meagring theory of designing that is innovative, radical, desirable, and inexpensive is articulated in **#37 A Swatchification Theory**. Likewise, **#68 A Jugaad Theory** of Design[ing] emphasises doing more with less by exhibiting the values of low-resource, makeshift, and unstructured innovation. Both **#69 A Gandhian Theory** and **#70 A Rasa Theory** of Design[ing] stress simplicity and principles that compassionately address the poorest of poor, society, nature and humanity at large. **#99 A Messy Theory** of Design[ing] highlights that design depends on the integration of both theory and practice and that theory, which is often invisible, is no less important in the way design is practised. Finally, **#113 A Post-Economic Theory** of Design[ing] censures design's dependence on the fiction of the economy as a domain separate from society and nature and advocates a praxis of design as commoning.

## Satiating

***[verb]***

*satisfied to the full; sated.*

Satiating theories of designing are those that satisfy to the full or even, perhaps, over-deliver and originate from the Latin satiatus, meaning 'enough', which prompts a few obvious questions. Has there already been there enough design(ing)? Is there enough design(ing) now? Will there ever be enough? That is, is enough enough or will enough ever be enough? Since design theory is generally invisible to design practice, it might be fair to ask whether for practice a theory can ever be satisfying? If as **#96 An**

**Excessive Theory** of Design[ing] claims that the average American home houses 300,000 things, then the practice of designing obviously gains its satisfaction from designing mountains of stuff making the practice of theory a luxury as **#01 A Luxurious Theory** of Design[ing] suggests. Then **#28 A Comfortable Theory** of Design[ing] points out we make sense of the world through our objects so a massive agglomeration of stuff leading to **#48 An Exaggerated Theory** of Design[ing] where normal is replaced by exaggerated designs to pursue pleasure. As such **#49 A Pornographic Theory** of Design[ing] contends we swim in an erotically designed semiosphere surrounded by the design of impossible offers - phew! It is timely that **#100 A Careful Theory** of Design[ing] urges design to embrace a life we care for.

# Instructing

***[verb]***

*give directions or information in order to tell or order someone to do something.*

That design is political before it is anything else is most evident in the manner in which it gives directions or information in order to guide or instruct someone to do something. Design is not political in the partisan sense as these instructing theories of design illustrate. What design thinks and how it does what it does do nothing but instruct us to do something (both for and against our will), sometimes fleetingly and sometimes for very long periods. Facing the consequences of automation **#20 An Automated Theory** of Design[ing] bids design to grow wild, while **#22 A Business Theory** of Design[ing] instructs us that being lazy might be better than being 'busy'. **#46 A FabLab Theory** of Design[ing] warns that we should remember that design is far from easy and therefore take greater care with it. So when a museum performs **#51 A Learning Theory** of Design[ing] it helps care for how we learn from object, discourse and performance, all of which can be practical. But **#53 A Practical Theory** of Design[ing] says we should rethink instruction and learn to learn as children do. **#61 An Educational Theory** of Design[ing] advocates a confrontational way to install a new point of view in education but warns this is a precarious condition. According to **#75 A Studio-based Theory** of Design[ing] what goes on in the studio has to address the cultural, moral, social and economic implications of projects. In that sense, does building an IKEA closet make you a designer or visiting a design 'specttacolo' make you a theoretician? #77 **An Inflation of Design Theory** of Design[ing] says no! According to **#87 A Constructivist Theory** of Design[ing], design might be considered the constructivist discipline par excellence, but it is also relativist and with relativism comes the risk of fundamentalism. In the same vein **#91 An Early-Learning Theory** asks us to find commonalities between design and science by starting with the fundamentals - A is for Art. **#98 A Design Thinking Theory** of Design[ing] advises design thinking is inseparable from design doing; in other words - designing. And **#107 A Designer-as-Mythologist Theory** of Design[ing] believes the designer should be a mythologist liaising with the world in sarcastic terms.

## Balancing

***[verb]***

*offsetting or comparing the value of one thing with another.*

For design to function, it performs a balancing act offsetting or comparing the value of one thing with another (see 'value analysis' which sets out to make a cheaper version of a competing product/service). But there exists a balancing act that causes uncertainty in design, and that is the critical comparison of theory with practice. This balancing act is characterised by hesitation between opposing considerations; weighing, pondering, compensating, comparing, or equalization, theory as distillation versus practice as derivation. The balancing theories of designing compare new variables with given systems. That we are at the halfway point between the beginning of life and the end is a fact, but according to **#12 A Cosmological Theory** of Design[ing], it should in practice incite new theories. Similarly, we are at the halfway point between the rural and urban populations, but very soon, three out of four people on the planet will live in cities. According to **#15 A Rural Theory** of Design[ing] we will compensate for this massive population imbalance through nostalgia. The rural now lives under a climate designed by capital in an atmosphere changed by design. **#25 A Climatic Theory** of Design[ing] can only really hope to salvage a better forecast for the strange weather ahead. In these turbulent conditions, **#62 An Evolutionary Theory** of

Design[ing] suggests that in the ideology of design the balance between the natural and the artificial is still evolving; and that leads directly to **#66 An Anthropocentric Theory** of Design[ing] where the theory is our anthropocenticity leads to our anthrochauvinism. As we boldly design more synthetic life for the increasingly artificial world **#93 A Catastrophic Theory** of Design[ing] exposes the early warning signs that synthetic life may actually represent a catastrophic theory of design. Climate change is throwing the weather out of balance, and **#104 A 1.5°C Theory** of Design[ing] warns with no coral reefs we should prepare for the waves. Then the following five theories shift their balance to relationships with land / gender / souths / the communal / the indigenous. **#105 A Cultural Generational Communal Iterative Generative Collaborative Relational Theory** of Design[ing] reclaims and claims generative, collaborative and respectful practices that give meaning to the agency of the more-than-human, the land and the relationships that connect everything. In a similar vein, **#110 A Communal Theory** of Design[ing] asserts the more we design the more we become the product of individualizing design forces when we should be designing for communal ways of being. The design of the south in **#114 An of the South(s) Theory** of Design[ing] brings together as many design equivalents as there are communities on Earth in what **#115 An Indigenous Theory** of Design[ing] calls a relational space that brings together entities in a Unity-that-is-all in which We-all-are-related. The disunity caused by design's problem-solving and diagrammatic binary relationships requires **#117 A Gendered Theory** of Design[ing] to support a feminist paradigm where design problems and solutions are co-emergent.

# Emoting

***[verb]***

*portraying emotion in a dramatic manner.*

Drama as we understand it was a gift of the ancient Greeks; first as tragedy where the gods did not joke with human folly, then as comedy where human folly did not need the gods to demonstrate its farcical existence (a history reprised by Slavoj Žižek in his 2009 book critical of the massive gift to banks following the financial crisis "First as Tragedy, Then as Farce").

The emoting theories of designing presented here dramatise the serious business of the dimension of design mostly missing from its histories – its emotional side. Where design is losing its substance and becoming invested with a quality of strangeness and remoteness **#16 A Romantic Theory** of Design[ing] asserts design is super-glueing happiness on everything. All romance has the potential for jealousy and according to **#32 A Jealous Theory** of Design[ing], jealousy, both good and bad, has long been associated with design. Jealousy demands faith and sentimentality is superficial. **#40 A Sentimental Theory** of Design[ing] warns design against indulging in the rational competitive frenzy; otherwise it risks becoming a sentimental practice. A model (of say - design theory) is the sum of all anecdotes and anecdotes (from the Greek) are things unpublished so when **#67 An Anecdotal Theory** of Design[ing] (or a theory about design theories) makes the case that design theory is ideological, anecdotal, and tautological this is worth publishing. Last, if design theory were to harness a broad array of theories, ranging from ancient philosophy, through haiku poems to social psychology, according to **#74 A Proud Theory** of Design[ing], it will create a new form of innovation.

# Dispiriting

***[verb]***

*causing someone or something to lose enthusiasm and hope; disheartening.*

It is difficult to decide which is the more dispiriting - the conviction powering 'what design can do' that harks back to the imaginary infinitely possible – or the systematic way design relieves itself of the infinite responsibility from doing what it does? Both fail to embrace the fact that our view of design reflects our view of society and this view is causing us to lose enthusiasm and hope. It is dispiriting, disheartening and depressing (most prescriptions are written to combat depression). And confronting the realities of designing in dark times might dim the glow of the possible but dispiriting theories of designing illuminate new possibilities such as **#30 A Matriarchal Theory** of Design[ing] that speculates whether persisting in the design of things that mothers used to do gives birth to a matriarchal theory of design. Another persistent current producing a very jaundiced theory of design is selling one's skill for very cheap, which according to

**#36 A Zombie Theory** of Design[ing] leaves design resembling a soulless corpse – which is the definition of a zombie. Green design generally credits design for appearing to be greenish, but **#43 A Green Theory** of Design[ing] sheds light on design's apparent greenness, immaturity and inexperience. These darkening times is the context for **#44 A Dark Theory** of Design[ing] that supposes that if design were to peer into how it has kept itself in the dark, then it might reveal something about its ignorance. By chronicling the pollution of the designed environment that humans cannot live without **#45 A Grey Theory** of Design[ing] aims to shake our faith in green ecology and intensify the blurry designerliness of design. Dispiritingly, instead of producing a better world **#47 A Barbaric Theory** of Design[ing] proposes design has been concerned with alleviating anxiety about consuming the world but to be a designer in a pandemic of inequality does not mean design has to be barbaric. But it does have to overcome the destructive endemic immaturity of technology, and **#57 A Mature Theory** of Design[ing] describes how design could contest our technological dependency. According to **#58 A Political Theory** of Design[ing], politics today is largely a fight between (privatized) profit and the interests of persons where design's brief is to explore the possibilities of creating societies in which accumulation is made secondary to an economy of needs. That's a powerful brief and **#60 A Power Theory** of Design[ing] maintains we need to work on expanding the consciousness of the contingent, precarious, limited character of power in design.

# Bounding

*[verb]*

*a limitation or restriction on action; place within certain limits; restrict.*

Bounding is the historic role of disciplines in which the boundary riders of the disciplines discipline the discipline. Bounding is also what we are discovering about the planet – it is bound in space-time and carrying capacity. It would appear that designing is coming to terms with the bounding of its practice as it was first enticed to flirt with stewardship then plunged headlong into sustainability to be rescued recently by resilience, all the while, finding comfort in limits. Within these bounds **#18 A Transitive Theory** of Design[ing] cautions design to seek beyond itself.

This is especially salient if, as **#55 A Data Theory** of Design[ing] warns, "design" is merely a shell for a database. If so, data must become the design and become intelligible to the user. To justify its shell-like structure **#56 A Justified Theory** of Design[ing] states design is justified in its own mind, rarely in that of others leading to circumstances that **#59 A Circumstantial Theory** of Design[ing] says seem limited. But designers who grasp circumstance open themselves to origination and all exemplary work in design follows this law. Escaping increasing categorization is a problem for design but **#80 An Everything and Nothing Theory** of Design[ing] helps escape the bounds when it says design is complex, messy, contradictory. In the construction of digital, networked, computational things design is not solid or stable and as such **#83 A Tuning Theory** of Design[ing] challenges us to develop practices of design as tuning formations of fluid assemblages that can harmonize relations of mutual integrity, meaningful transparency, and democratic innovation. The digital is also bringing about the emergence of a new kind of design replacing centuries-old Cartesian principles and as reasoned by **#90 A Non-Cartesian Theory** of Design[ing] in this scenario scale and standards are simply vanishing. If this is so then according to **#92 A Scientific Theory** of Design[ing] it's plausible to speculate upon a theory of design that is boundless and pre-disciplinary, just like science. And if this sounds fictional then as **#95 A Fictional Theory** of Design[ing] reminds us, in the era of the production of nothing design is producing fictions only by evading the truth. In this make-believe setting what better way to restore the boundaries than **#97 A Management Theory** of Design[ing] which illustrates how management seemed to be the ideal way in which to bring discipline to an undisciplined discipline.

## Bettering

***[verb]***

*improve on or surpass (an existing or previous level or achievement); make something better; improve.*

As these theories illustrate making something better hasn't been smooth sailing for design. Despite strident warnings from Victor Papanek in his 1971 book 'Design for the Real World', and design's own mantra about bettering the world, rather than make the world better design has been content to improve an existing or previous level or achievement. What bettering theories of designing illustrate is how hollow that mantra rings. In fact, bettering by design more closely resembles the legal definition of betterment – to enhance the value of real estate – the commodity of Capital par excellence. The sum of all possibilities is equal to the sum of all responsibilities for what design imagines is possible according to **#04 A Possible Theory** of Design[ing]. As such **#07 A Future Theory** of Design[ing] asks will we be automated or unemployed, unequal or underwater, radicalised or tranquilised…for sure the future that is traded is derived from uncertainty, while the future that is predicted is certain never to come true. One thing for sure **#13 An Ultimate Theory** of Design[ing] questions whether life and death are now just more consumer choices and if so, then design will need to consider its ultimate theory. And **#14 A Utopian Theory** of Design[ing] claims the very idea of design is already utopian because, for a brief moment in history, it seemed design was capable of answering the question – what-might-become? If we could answer that, then **#35 A Winning Theory** of Design[ing] explains how, by design, TV made winners from the failure of others and, by design, any failure can still result in a win for design. If that isn't a fantastic prospect **#41 A Fantastic Theory** of Design[ing] explains how design plays an imaginary game of making things better through its fantasy of talent, of creativity, of innovation, and more recently of disruption while studiously avoiding ideology. **#78 An Idealist Theory** of Design[ing] makes it clear that without ideology, we can't expect care and empathy from designers (but we can all be branded as ideological capitalist pigs). The prime objective of **#85 A Speculative Theory** of Design[ing] is to force an aspect of the future into the present so that it demands a response in the form of possible future scenarios to illustrate the urgency of change required. A tricky balancing act is posed by **#86 A Global Health Theory** of Design[ing] which points out that designers are being asked to promote healthy lifestyles plus help reduce the cost and burden of disease. **#88 A Caring Theory** of Design[ing] and practice intra-acts to make a difference in the world. So if we acknowledge the ever-increasing contradictions in which design operates **#101 An Agile Theory** of Design[ing] describes how it becomes more and more impossible to consider design as static. The SMART paradigm landed on design from the corporate tech world and appears to be imperative for contemporary designers, but as **#103 A SMART Theory** of Design[ing] indicates, it isn't making things much better. **#111 An Autonomous Theory** of Design[ing] asks whether we could fathom the case of small bands of designers exercising their craft in support of communities in struggle? The answer - design that strengthens the collective autonomy of communities is autonomous design. And since design has to now do what it can in the face of what-might-not-become **#118 A Better Theory** of Design[ing] cautions design to face up to what it cannot do.

## Galvanising

***[verb]***

*shock or excite (something or someone) into taking action.*

When Luigi Galvani stimulated frog legs with electricity, he gave us a new form of drive – a verb – galvanising – meaning shocking or exciting something or someone into taking action. While it tends to shy away from its past and dismiss theory, design has never been shy of taking action. Without doubt, design would like to promote itself as a galvanising practice (how else to change everything), but this is at odds with the discipline's virtuous trend to participatory and cooperative design. While practice equivocates, theory is galvanising as these theories show. Many significant designers have tried to remind us - perhaps too long ago for most memories - that design is a serious business. **#23 A Responsive Theory** of Design[ing] points out we will have some challenges responding. According to **#24 A Sharing Theory** of Design[ing] we will struggle to reconcile design's consenting indebtedness to capitalism with its rhetoric about caring for the one world we share. If we were to ask whether or not anonymity today is a designer's biggest ambition or their greatest phobia **#27 An Anonymous Theory** of Design[ing] proposes that it would be interesting to know how many of today's superstar designers would be happy to remain anonymous. The history of the distinction between practical subjects and theoretical subjects, between design and theory, between doing and thinking leads **#52 A Theoretical Theory** of Design[ing], which asks whether we can distinguish between a theory or a practice lesson. **#54 Against a Theory** of Design[ing] warns against the dichotomy between doing well and thinking well. Choosing between a solid knowledge of design practice versus a practice of a theory can tell us a lot about our idea of design. In pluralistic society, everyone is different from oneself. In the same way design's plurality is in itself, not in its products. Rather than confuse **#71 A Pluralistically Confusing Theory** of Design[ing] resolves how plurality enriches design. According to Dieter Rams "*...indifference towards people and the reality in which they live is actually the one and only cardinal sin in design.*" **#76 An Inclusive Theory** of Design[ing] asks can we trust designers to consider us as the fantastic automated and seamless world is being materialised? Is it possible to design in a vacuum with no sense of who might use your designs, where they might be used or what impact their use might have on others or the environment? **#81 An Immersive Theory** of Design[ing] doesn't think so. We use the 'lightbulb moment' as a metaphor for inspiration. **#82 A Failing Forward Theory** of Design[ing] reminds us it apparently took Edison 1000 failures to perfect a working light bulb so to design is to fail. The gaming of everything by design might be booming but **#84 A Ludic Theory** of Design[ing] reminds us that because play drives our development as humans, research takes play very seriously and so too must design. Who/what are the animate/inanimate, natural/technological agents of more-than-human design asks **#89 A More than Human(ist) Theory** of Design[ing] and how might they be engaged in design processes and experiences? Another **#102 A Ludic Theory** of Design[ing] supposes that since playfulness is a vital component of creativity, all design must be playful at some stage. Not so playful is **#106 A Plurality Theory** of Design[ing], which is concerned with future thinking to seed change and modify power structures. If life is always ceaseless flow, endlessly reverberating into milliard forms, then **#112 A Pluriversal Theory** of Design[ing] proposes a world where many worlds fit. Enacting a declassified design, intentionally reversible and provisional, requires the de-design of any classificatory order. According to **#116 A Declassified Theory** of Design[ing] declassifying is not opposed to classifying; rather it entails reclassifying with other knowledges and purposes.

# 42 Destabilising

*[verb]*

*upset the stability of (a product or system); cause unrest or instability.*

If the mission of design is some sort of futuring, then the ground from which it projects futures has to be stable. Stable ground was a necessary condition for modernisation, but flows of money, goods and people have collided with planetary limits and upset the stability of products and systems; causing unrest and instability. What the destabilising theories of designing reveal is that design, having destabilised the human orbit with an excess of things, is itself now being destabilised by its history of excess. Once duplication stipulated maximum simplicity but **#02 A Duplicate Theory** of Design[ing] proposes that in the hyper-infinite digital production process, infinite customisation is mandatory and duplication is unnecessary. The aim of **#05 A Disruptive Theory** of Design[ing] is to creatively rebalance the artificial world using a (well designed) spanner to challenge existing social, political, cultural, economic, environmental and technological norms. The fragmentation of disciplines has spawned irresponsible and undisciplined designers who ignore distinctions and help themselves to methods. **#06 An Undisciplined Theory** of Design[ing] celebrates disorder, disruption and rebellion. There is hardly anything in our daily lives that has not been designed, and as a result **#09 An Imperfect Theory** of Design[ing] explains that part of what makes us human is everything is imperfect. Far from perfect is the global domination of Capital under which the future is being traded as the 'derivative'. **#11 A Derived Theory** of Design[ing] states all design action now becomes something we will pay for in the future. Even though every design problem begins with an effort to achieve fitness between form and context, wrongness is more compelling. Accordingly, **#26 A Misfit Theory** of Design[ing] embraces wrongness in all its guises. Understanding how little we know about design is paradoxical but not destabilising. **#29 A Paradoxical Theory** of Design[ing] invites us to recognise that the unknown is a necessary and welcome condition for beginning all design. There is so much that is alluring and dangerous to young designers warns **#31 A Dangerous Theory** of Design[ing] especially the desperation to scoop some visibility and disregarding this temptation is a crime. When design can be anything – from drugs to software – "mongrel" institutions will flourish, and in this postmodern soup **#33 A Mongrel Theory** of

Design[ing] forewarns design can also be nothing. Having exceeded the planet's carrying capacity and completely destabilised it, **#34 A Tired Theory** of Design[ing] is a timely reminder that our future will not be driven by energy, accumulation, and expansion, but by the design of slowness, withdrawal, and tiredness. Because so much news is fabricated false news can be hard to distinguish, but as **#39 A False Theory** of Design[ing] illustrates, design is complicit meaning a false theory of design is no joke. Plastic and design have a long, carefree and productive history that has filled up our homes, landfill and oceans, and plastic made design look creative. **#42 A Plastic Theory** of Design[ing] asks, if design were truly plastic, it would exhibit the ability to adapt to planetary change. If having to be plastic isn't tough enough, according to **#73 A Bipolar Theory** of Design[ing], design suffers from inferiority regarding its image, exaggerated prestige, and struggles to define its place in the world making it easy prey for business. Design/ing creates worlds and design/ers fashion ways of being, thinking and doing. But **#94 A Disempowering Theory** of Design[ing] asks whether the so-called "dark patterns" of interface design should gently guide users toward the light of intended behaviour or should they not? While this question is not answered, **#108 An Ontological Theory** of Design[ing] serves notice, that by designing, we are designed. So who/how shall we build the prototypes of the multiple civilizations of the futures to come? From these extremes the most destabilised relationship is the Dream of the Earth about which **#109 A Materials' History Theory** of Design[ing] asks what do plastics - or so many new materials - know?

## Deflating

***[verb]***

*reduce the level of (an emotion or feeling).*

Deflating is the reverse of inflating and design has been inflating its recognition for too long; for example, the soaring media coverage of what designing is doing is inversely proportional to its relevance. After inflating the practical relevance of design, it might be the job of theory to deflate it (just a little), so it might better counteract the lopsided inflation of design's worth. This batch of theories sets about deflating the self-pronounced importance of design in various ways. **#17 A Radical Theory** of Design[ing] advocates design act in the manner of the verb – to radical – that is radicalling our thinking before reminiscing about the inflated relevance of radical design. Design is comfortable with its known knowns. It avoids the known unknowns while ignoring the unknown unknowns. But **#19 An Unknown Theory** of Design[ing] describes the well known unknown – the better world – which is better only for those who already own it. Anti-design was motivated by the dwindling relevance of design. **#38 An Anti-design Theory** of Design[ing] reminds us that design's relevance resides in challenging the status quo, confronting convention and opposing commonly held assumptions. Admitting that our world is a troubled place, **#50 A Curious Theory** of Design[ing] believes in curiosity as the basis of design action and that even serious and controversial problems are better addressed with a sense of humour. So it is funny when **#63 An Uncertain Theory** of Design[ing] speculates that the only thing that might be certain about design is that sometimes our design intentions might not quite turn out as intended. According to **#64 A Pessimistic Theory** of Design[ing] a major unintended consequence of our propensity to design is – we may well design ourselves right out of existence. Since design thinking is widely regarded as key to addressing our most complex challenges but implicated in the conception of excess **#65 A Thoughtless Theory** of Design[ing] wonders whether we have lost the imagination to think differently. Following the social, political and economic turmoil of the last decade, **#72 An Anti-Design Thinking Theory** of Design[ing] suggests a true and significant innovative turn stems from just good old multi-disciplined... well – thinking. If we are not going to start abandoning design theories that do not offer us a pluralistic, situated and relational view of design then **#79 An Irrelevant Theory** of Design[ing] asks why have theories at all if they are flawed and unfit for the changes ahead?

#01

# Lu

Luxurious

# #01
# A Luxurious Theory of Design[ing]

*Paul Rodgers & Craig Bremner*

Much awarded design activity is aimed at luxury, whether in the obvious things of luxury or luxury serviced as heightened forms of pampering. The word luxury is derived from the Latin for excess and excess is today a precondition of lifestyle for an increasing quantity of people. It augments the so-called necessities of life, making requisite the not indispensable, and what is already choice and costly, into lascivious avocations. From Freeports concealing irrationally costly objects, so their trade avoids tax to ports harbouring ridiculously conspicuous vessels of tax avoidance, design plays an increasingly flagrant role. A luxurious theory of design specifies the laws of excess in which design is engaged to intensify the enjoyment of already indulgent products and services. In this element, it is a luxury for design to entertain itself with theory at all.

### Notes and Further Reading

Neil Fitzgerald, "Speculative Aesthetics: Freeports as the Art Caves of High Finance", Verso Books, 28 November 2016, http://www. versobooks.com/blogs/2964-speculative-aesthetics-freeports-as-the- art-caves-of-high-finance

Robert H. Frank, *Luxury Fever: Money and Happiness in an Era of Excess*, Princeton, NJ: Princeton University Press, 2000.

Lambert Wiesing, *A Philosophy of Luxury*, London: Routledge, 2019.

#02

# Du

Duplicate

# #02
# A Duplicate Theory of Design[ing]

*Paul Rodgers & Craig Bremner*

Complex objects serving complex functions do not require a designer. How can this be? Darwin's theory of evolution advocates that complexity grows through systematic duplication and inevitable accidents in duplication eventually produce what over time appear to be new designs for duplication. In the machine production process, infinite systematic duplication was mandatory since mass production was the most efficient route to the lowest price. Now in the hyper-infinite digital production process, infinite customisation is mandatory, and duplication is unnecessary. In this scenario, every service becomes something akin to an accident but does not produce a new efficiency. In fact, it produces maximum complexity about the terms and conditions for the service. Amazon's Kindle e-reader contract is 73,198 words long and would take the average person nine hours to read. A duplicate theory of design makes complex functions of simple objects.

### Notes and Further Reading

Michel Foucault, *The Order of Things: An Archaeology of the Human Sciences*. New York, NY: Vintage Books, 1994.

Rebecca Goldstein, *36 Arguments for the Existence of God*. London: Atlantic Books, 2010.

Adam Greenfield, Radical Technologies: The Design of Everyday Life, London: Verso Books, 2017.

#03

# Af

Affordable

# #03
# An Affordable Theory of Design[ing]

*Paul Rodgers & Craig Bremner*

Design borrowed the idiom of affordance to prescribe how we might be persuaded to use the world as it has been designed. In the same vein, design has always claimed with each act of design that the world will be used better. The concept of the better world is one of the enduring myths of design. Design invested in affordances on the basis that usability is essential for both profit and pleasure. But the investment in affordance runs counter to the obvious fact we don't use the world as it was designed to be used. We use it to suit our own designs. Had design paid attention to the verb – to afford – it would have discovered its real investment – *i.e.* furthering a world that is becoming increasingly unaffordable to anyone who doesn't already own it. An affordable theory of design might not be usable, but it is fashionable.

## Notes and Further Reading

James J. Gibson, *The Senses Considered as Perceptual Systems*, London, UK: Allen and Unwin, 1966.

James J. Gibson, "Affordances and Behavior", *In* E. S. Reed & R. Jones (Eds.), *Reasons for Realism: Selected Essays of James J. Gibson*, Hillsdale, NJ: Lawrence Erlbaum, 1975 (1st edition), pp. 410 - 411.

James J. Gibson, *The Ecological Approach to Visual Perception*, Boston, MA: Houghton Mifflin Harcourt (HMH), 1979.

Don Norman, *The Design of Everyday Things*, New York, NY: Basic Books, 2013 (Revised and Expanded Edition).

#04

# Po

Possible

# #04
# A Possible Theory of Design[ing]

*Paul Rodgers & Craig Bremner*

Michel Foucault stated that he didn't "*...write a book so that it will be the final word; I write a book so that other books are possible, not necessarily written by me.*" (O'Farrell, 2005: 9). If that is a reasonable explanation for the act of writing then a reasonable paraphrase for the act of designing might read the aim of designing possible future scenarios is not to prescribe the conditions of the scenario nor corral the future but to ensure the possibility of the possible – *i.e.* only designing makes design possible. The role of theory in this possibility is to explain why the world looks like it looks. The role of technology in this possibility is to illustrate infinite possibilities. The sum of all possibilities is equal to the sum of all responsibilities for what design imagines is possible.

## Notes and Further Reading

Nicolas Bourriaud, *The Exform*, London: Verso Books, 2016 (translated by Erik Butler).

Michel Foucault, "Entretien avec Michel Foucault." in *Dits et Ecrits vol II.* Paris: Gallimard, 1994 [1971], pp. 157 - 174.

Boris Groys, Self-Design and Aesthetic Responsibility, *e-flux Journal*, No. 7, 2009.

Alexandra Midal, *Design by Accident: For a New History of Design*, Berlin: Sternberg Press, 2019.

Clare O'Farrell, *Michel Foucault*, London, UK: Sage Publications, 2005.

#05

# Di

Disruptive

# #05
# A Disruptive Theory of Design[ing]

*Paul Rodgers & Craig Bremner*

A disruptive theory of design[ing] is not about being unruly, badly behaved, rowdy, and unmanageable. It is, however, about being undisciplined, playing havoc with the status quo and contesting the ways that things are usually done. It is about disturbing and interfering with conventional mindsets and conventional processes and policies. A disruptive theory of design aims to upset, unsettle, and interrupt things. Metaphorically speaking, the aim is to throw a (well designed) spanner into the worlds of policymakers, management, governments, commerce, industry and others who are unwilling to shift from a world that is unequal and out of kilter on many levels. A disruptive theory of design is about creatively rebalancing the artificial world by challenging existing social, political, cultural, economic, environmental and technological norms.

**Notes and Further Reading**

Alastair Fuad-Luke, *Design Activism: Beautiful Strangeness for a Sustainable World*, London, UK: Routledge, 2009.

Paul Paetz, *Disruption by Design: How to Create Products that Disrupt and then Dominate Markets*, New York, NY: Apress, 2014.

#06

# Un

Undisciplined

# #06
# An Undisciplined Theory of Design[ing]

## *Paul Rodgers & Craig Bremner*

An undisciplined theory of design[ing] is exactly that. Undisciplined. But rather than seeing undisciplined as being something negative and akin to badly behaved, naughty, and even out of control, one should celebrate the disorder, disruption and rebellious nature of undisciplined design. Undisciplined designers see no walls between creative terrains. They shape future visions without distinguishing between one discipline and another. This emerging creative eclecticism is at the heart of this new undisciplined way of doing things, and these new undisciplined designers are the real harbingers of change. The fragmentation of distinct disciplines, including those located in a sea of traditional art and design contexts, has altered creative practice from being 'discipline-based' to 'issue- or project-based' (Heppell, 2006), which emphasises and empowers irresponsible and undisciplined creative practitioners, who purposely ignore distinctions and who borrow and utilise methods, techniques and tools from many places.

### Notes and Further Reading

Craig Bremner and Paul A. Rodgers, "Design without Discipline", *Design Issues*, Vol. 29, Issue 3, Summer 2013, pp. 4 – 13.

Stephen Heppell, RSA lectures: Stephen Heppell – learning 2006 [Video file]. Available from: http://archive.teachfind.com/ttv/www.teachers.tv/videos/stephen-heppelllearning-2016.html

Paul A. Rodgers and Craig Bremner, *Design School: After Boundaries and Disciplines*, Malaga, Spain: Vernon Press, 2019.

#07

# Fu

Future

# #07
# A Future Theory of Design[ing]

*Paul Rodgers & Craig Bremner*

A future theory of design is very different from a derived theory of design, although both trade the future in the mercantile spectacular. The future that is traded is derived from uncertainty while the future that is predicted is certain never to come true. Despite this fact, the engine of capitalism has organized us to invest in our joyous future through the product called a pension. Meanwhile, the past is now a product in more demand than the future with so many reactionary political platforms being built on the future revival of the past. Between pensioning certainty and profiting from uncertainty, there are probably countless contradictory predictions about the future. Will we be automated or unemployed, unequal or underwater, radicalised or tranquilised, rentier or rentism (look that one up – Frase, 2016). For sure, whatever will be will be the future theory of design[ing].

### Notes and Further Reading

Franco "Bifo" Berardi, *Futurability: The Age of Impotence and the Horizon of Possibility*, London, UK: Verso Books, 2017.

Peter Frase, *Four Futures: Life after Capitalism*. London, UK: Verso Books, 2016.

McKenzie Wark, *Inventing the Future*, Public Seminar, October 27, 2015. Available at http://www.publicseminar.org/2015/10/inventing-the-future/.

R. John Williams, "World Futures", *Critical Inquiry*, Vol. 42, No. 3, 2016, pp. 473 - 546.

#08

# Ch

Cheap

# #08
# A Cheap Theory of Design[ing]

*Paul Rodgers & Craig Bremner*

Backed by Google Ventures, VIDA boasts that it is "*...a platform where artists can adapt their artwork into fashion products by converting their original artwork into elegant apparel and accessories.*" If the subject is born of the object as Michel Serres asserts, then VIDA and 'Maker-This' or 'Maker-That' supporting the production of stuff by increasingly cheapening 3D printers to be sold cheaply through Etsy or other micro-business platforms, would indicate that the contemporary subject is just a cheap (and cheerful) accessory. If there is nothing beneath appearance, then two lines from a recent Leonard Cohen song reveal a totally overlooked theory of design. He sings - "*As he died to make men holy / Let us die to make things cheap*" reminding us that together with a cheapening theory of design, in practice design has always been dying to make things cheap.

## Notes and Further Reading

Leonard Cohen, "Steer Your Way", from *You Want It Darker*, Columbia Records, 2016.

Raj Patel and Jason W. Moore, *A History of the World in Seven Cheap Things: A Guide to Capitalism, Nature, and the Future of the Planet*, Berkeley, CA: University of California Press, 2018.

Michel Serres, *Statues: Le second Livre de Fondations*, Paris: Flammarion, 1987.

VIDA - https://triplebyte.com/company/public/vida-co

#09

# Im

Imperfect

# #09
# An Imperfect Theory of Design[ing]

*Paul Rodgers & Craig Bremner*

Recognising and dealing with imperfection is part of what makes us human. Everything is imperfect. As David Pye so eloquently articulated in his seminal work, 'The Nature and Aesthetics of Design', "*Nothing we design or make ever really works. We can always say what it ought to do, but that it never does.*" (Pye, 1978: 13-14). So, striving for design perfection is a pointless pursuit. Every day of our lives, we are exposed to a vast array of products, systems, services, and spaces. There is hardly anything in our daily lives that has not been designed. However, most of these useful things do useless things that no one wants them to do. Trains frequently breakdown, our computers crash regularly, our dinner table should be impervious to scratches and be self-cleaning – but they are not of course, and our refrigerators, air conditioning units, and homes all consume valuable natural resources like a hungry animal.

## Notes and Further Reading

David Pye, *The Nature and Aesthetics of Design*, London, UK: The Herbert Press, 1978.

Davis Pye, *The Nature and Art of Workmanship*, London, UK: The Herbert Press, 2007.

Robert Stadler and Alexis Vaillant (Editors), *On Things as Ideas*, Berlin, Sternberg Press, 2017.

#10

# Hu

Humble

# #10
# A Humble Theory of Design[ing]

*Paul Rodgers & Craig Bremner*

Designers need to show more humility. Although many of them are quite brilliant, talented, and know loads about lots of stuff, they do not have all of the answers. Nor should they. Designers are at their best when they work with, empower and rely on others. Designers are experts at bringing people and other resources together in powerful ensembles. Design is never a solo pursuit. Design, by definition and necessity, involves a vast array of actors – clients, commissioners, manufacturers, salespeople, and many others. Acknowledging that one might know very little about many things is the first step towards greater enlightenment for design practice. Starting from a perspective of complete ignorance requires one to embrace unfamiliarity, inexperience, cluelessness, and an honest lack of knowledge that will be totally refreshing to all involved.

### Notes and Further Reading

Kenya Hara, *Designing Design*, Zurich, Switzerland: Lars Muller Verlag, 2007.

Kenya Hara, *Ex-formation*, Zurich, Switzerland: Lars Muller Publishers, 2015.

Jacques Ranciere, *The Ignorant Schoolmaster: Five Lessons in Intellectual Emancipation*, California, USA: Stanford University Press, 1991.

#11

# De

Derived

# #11
# A Derived Theory of Design[ing]

*Paul Rodgers & Craig Bremner*

After exhausting the carrying capacity of the planet, the one resource left to the project of Capital is the future. As the future is being traded as the 'derivative' under the global domination of Capital, the possible future scenarios that design was once called upon to project have changed form and function and now derive their currency from the derivative. Derivative markets trade in risk and uncertainty. Possible futures become traded as financial goods and, in that process, derivatives create 'goods' out of 'bads'. In theories of risk and uncertainty, risks are held to be future, yet real, events. They do not exist as yet, but they nevertheless inform the present. With the objectification of uncertainty, the payment for risk has been conveniently shifted to future generations. A derived theory of design means all design action now becomes something we will pay for in the future.

### Notes and Further Reading

Arjun Appadurai, *Banking on Words: The Failure of Language in the Age of Derivative Finance*, Chicago, IL: University of Chicago Press, 2015.

Jakob Arnoldi, "Derivatives: Virtual Values and Real Risks", *Theory, Culture and Society*, Vol. 21, Issue 6, 2004, pp. 23 - 42.

Adam Arvidsson, "Facebook and Finance: On the Social Logic of the Derivative", *Theory Culture and Society*, Vol. 33, Issue 6, 2016, pp. 3 - 23.

#12

# Co

Cosmological

# #12
# A Cosmological Theory of Design[ing]

*Paul Rodgers & Craig Bremner*

In describing the two bookends of modern science, Jean-Francois Lyotard points out that the weak signals we can detect from the Big Bang resonate with the prospect that we might eventually 'witness' our birth. Plus, he reminds us that because the sun is at the halfway mark of its life span, with its death, we can foresee our own death. For Lyotard, the impending death of all humanity necessitates the birth of new ways for thought to go on without a body. Every design project, too, faces a cosmological crisis because, despite its best intentions, the project for a better world is probably dead. In fact, it was never really born. To put in motion a cosmological theory, design will have to change orbit from satellite of capital to augury in the birth and death of the planet.

## Notes and Further Reading

Debbora Battaglia, "Cosmic Exo-Surprise, or, When the Sky is (Really) Falling, What's the Media to Do?", *e-flux Journal*, No. 46, June 2013.

Jean-François Lyotard, *The Inhuman: Reflections on Time*, Cambridge: Polity Press, 1991.

Siegfried Zielinski, Peter Weibel, and Daniel Irrgang (Editors), *Flusseriana: An Intellectual Toolbox*, Minneapolis, MN: Univocal Publishing, 2015.

#13

# Ul

Ultimate

# #13
# An Ultimate Theory of Design[ing]

*Craig Bremner & Paul Rodgers*

Towards the end of Radical Design, Alessandro Mendini wrote "The Utility of Uselessness" (Mendini, 1979) exploring the ironic difference between useful and useless. In the same vein, he also designed a cast bronze briefcase for our 'final journey' - death. Superficially useless but useful as a memento mori, Mendini's modern symbol serves to remind us that death is also being turned into a consumerist choice under the brand 'voluntary euthanasia'. The briefcase and the brand are useful prompts to questions about how will we be allowed to live or die, by what criteria, and who gets to choose? It is useful to remember that in the recent past 'voluntary' became involuntary, turning the design for a 'well-death' into designs for death. If life and death are now just more consumer choices, then design will need to consider its ultimate theory.

### Notes and Further Reading

Alessandro Mendini, "Valigia per l'ultimo viaggio", *Braccio di ferro per Cassina*, 1972.

Alessandro Mendini, *Utilità dell'inutile* (*The Utility of Uselessness*), Modo 21, 1979.

James Pallister, "Reinventing Death for the Twenty-first Century", Design Council, 5 May 2015, Available at: http://www.designcouncil.org.uk/news-opinion/reinventing-death-twenty-first-century-0?utm_source=Design%2BCouncil%2BNewsletter&utm_medium=email&utm_campaign=newsletter_150505

#14

# Ut

Utopian

# #14
# A Utopian Theory of Design[ing]

*Craig Bremner & Paul Rodgers*

A utopian theory of design is a tautology. After all, the very idea of design is already utopian. Why? Because when idea split from manufacture design became an idea in search of an ideal location – the island of "what next?" Design was bequeathed a warrant to imagine change at the same time as change was taking place, simply because it could be imagined. Imagining this revolution to be the project of the better world, design imagined that it was capable of answering the question – "what-might-become?" Hence the tinge of utopian hope. However, design failed to notice the machine had instantly turned the imagination of change into an image of itself turning everything into a derivative. Now the machine has been replaced by Capital rapidly consuming the future design is stranded on an imaginary island searching for a rationale for "what-might-not-become?"

## Notes and Further Reading

Zygmunt Bauman, *Retrotopia*, Cambridge: Polity Press, 2017.

Sandra Groll (Translated by Kate Hunter), "Traces and Hopes of Design Research: An Interview with Gui Bonsiepe, Klaus Krippendorff, Siegfried Maser, and René Spitz", *Design Issues*, Vol. 31, No. 1, Winter 2015, pp. 18 - 31.

Marshall McLuhan, *Understanding Media: The Extensions of Man*, New York, NY: McGraw Hill, 1964.

China Miéville, "The Limits of Utopia", *Salvage* #1, July 2015. Available at: http://salvage.zone/in-print/the-limits-of-utopia/

Thomas More, Utopia, London, UK: Verso Publishers, 2016.

John Wood, Design for Micro-Utopias: Making the Unthinkable Possible, London, UK: Gower Publishers, 2007.

#15

# Ru

Rural

# #15
# A Rural Theory of Design[ing]

*Craig Bremner & Paul Rodgers*

According to the United Nations, on a day in 2007, the population of the planet flipped from rural to urban. By the end of the 21st century, the urban will swallow up 75% of the world's population (with 60% in slums). Despite the pull of the urban crisis on design, the future of the rural remains consequential for two reasons. The first trading with the future was gambling with the price of farm produce from which we learnt to gamble with the planet's future – not its forward price but its potential design for profit before it is exhausted. And long before the mass-production of city-living, rural-living provided an infinitely customisable design for being together in increasing number and proximity. While we speculate on living in massive cities, nostalgia for what is not urban will drive future demand for a rural theory of design.

### Notes and Further Reading

AMO / Rem Koolhaas, *Countryside: A Report*, Cologne: Taschen, 2020.

Vanessa Miriam Carlow (Editor), *Ruralism: The Future of Villages and Small Towns in an Urbanizing World*, Berlin: Jovis, 2016.

Jeremy Davies, "Sustainable Nostalgia", *Memory Studies*, Vol. 3, Issue 3, 2010, pp. 262 - 268.

The World Bank, Transport and Urban Development Department, "Report on the Dynamics of Global Urban Expansion", 2005, Available at: http://web.worldbank.org/WBSITE/EXTERNAL/TOPICS/EXTURBANDEVELOPMENT/0,,contentMDK:20970341~pagePK:148956~piPK:216618~theSitePK:337178,00.html

#16

# Ro

Romantic

# #16
# A Romantic Theory of Design[ing]

*Craig Bremner & Paul Rodgers*

According to the nineteenth-century American writer Nathaniel Hawthorne, where things seem to lose their actual substance and become things invested with a quality of strangeness and remoteness, this illustrates the inner-workings of the genre called Romance. In the era of production without a product, this is a fairly accurate description of the contemporary genre of service design. Or, in the era of self-design, this is a fairly accurate description of the genre of self-production. The aim of self-production is to superimpose character and personality with appearance. The aim of self-design is to superpose appearance with image. In service design, the actual substance of things is superseded by images invested with self-satisfying and optimistic promises of happiness. In this scenario, design better resembles the genre of the romance novel. A romantic theory of design is where every image superglues happiness on everything.

## Notes and Further Reading

Franco "Bifo" Berardi, "Engineering Self", *e-flux Architecture*, December 2016, Available at: https://www.e-flux.com/architecture/superhumanity/66877/engineering-self/

Boris Groys, The Obligation to Self-Design, e-flux Journal, #00, November 2008.

Nathaniel Hawthorne, *The Scarlet Letter: A Romance*, Boston, MA: Ticknor and Fields, 1850.

Geert Lovink, *Sad by Design: On Platform Nihilism*, London, UK: Pluto Press, 2019.

#17

# Ra

Radical

# #17
# A Radical Theory of Design[ing]

*Craig Bremner & Paul Rodgers*

Mention of a radical theory of design[ing] probably immediately excites in most designers thoughts about the influential period of radical design that Andrea Branzi of Archizoom called the long decade in a short century. And let's not quibble – radical design was radical. Increasingly, designers look back at radical design, wondering why it is no longer possible to reprise the spirit of this age. The expanding portfolio of attempts to repair design with the radical have only simply been designated strange design. A radical theory of design[ing] would not restore the noun radical as an essential element of design, nor repurpose radical as an adjective in the hope of reanimating design itself. Instead, a radical theory of design[ing] would advocate design acts in the manner of the verb – to radical – that is radicalling our thinking before reminiscing about radical design.

## Notes and Further Reading

Jehanne Dautry and Emanuele Quinz (Eds), *Strange Design: From Object to Behaviours*, France: it: Editions, 2015.

Lizzie O'Shea, *Future Histories: What Ada Lovelace, Tom Paine, and the Paris Commune Can Teach Us about Digital Technology*, London, UK: Verso Books, 2019.

Marco Petroni, *Mondi Possibili: Apuntes de Teoría del Diseño*, Milano, Italy: Edizioni Temporale, 2016.

#18

# Tr

Transitive

# #18
# A Transitive Theory of Design[ing]

*Paul Rodgers & Craig Bremner*

As if it needs a distinctive feature to prop it up, design always seems to be labelled with an adjective. Inclusive, speculative, critical, adversarial, fictional, interactive, discursive, interrogative... The variety of traits is long. Recently, transition has been affixed to bolster design, and it comes at a moment when the prefix 'trans' no longer prefigures making a difference but foreshadows the need to reassign difference. Whereas transformation promised something completely different, transition design at best can only manage to reorganise things. As such, transition will not transform design. Given a central promise of design is the passage to a better future, the trouble with transition is that it resembles a transit lounge where, awaiting the next leg of the passage back to the familiar, design is becoming increasingly intransigent. In this situation, a transitive theory of design cautions design to seek to see something beyond itself.

## Notes and Further Reading

Clino Trini Castelli, *Transitive Design*, Milan: Mondadori Electa, 2000.

Anne-Marie Willis, "Transition Design: The Need to Refuse Discipline and Transcend Instrumentalism", *Design Philosophy Papers*, 13:1, 2015, pp. 69-74.

#19

# Un

Unknown

# #19
# An Unknown Theory of Design[ing]

*Craig Bremner & Paul Rodgers*

When Donald Rumsfeld made a now-famous statement saying that something that hasn't happened is always interesting to him, because as we know, there are known knowns – things we know we know, known unknowns - we know there are some things we do not know, and unknown unknowns - things we don't know we don't know. Slavoj Žižek extrapolated from Rumsfeld's three categories a fourth - the unknown known - things we intentionally refuse to acknowledge that we know (Žižek, 2004). What do these statements mean for design? First, there are the known knowns where the disciplines control the disciplines producing objects that consumers buy. Second, there are the known unknowns where design knows it needs to go but is doing everything it can to make sure no-one goes there. Third, there are the unknown unknowns where design has been unknowingly coated with unknowable layers like sustainability and ethics and responsibility. Last is the unknown known - the unassailable modern project for a better world that we know is only better for those who already have everything. We don't want to know that design is engaged in the design of inequality, of weapons, of war, of surveillance networks, and so on.

### Notes and Further Reading

Kenya Hara, *Designing Design*. Zurich: Lars Muller Verlag, 2007.

Kenya Hara, *Ex-Formation*, Zurich: Lars Muller Publishers, 2015.

Donald Rumsfeld, Department of Defense news briefing, 12 February 2002.

Slavoj Žižek, "What Rumsfeld Doesn't Know that He Knows About Abu Ghraib." In These Times, 21 May, 2004. http://inthesetimes.com/article/747/what_rumsfeld_doesn_know_that_he_knows_about_abu_ghraib

#20

# Au

Automated

# #20
# An Automated Theory of Design[ing]

*Craig Bremner & Paul Rodgers*

Automation is gaining on us as we automate everything to automatic equipment, systems, or processes. The prospect of the apparently inevitable and rapid automation of all work currently done by humans is well underway. The benign ATM is a good example of the strategy – a convenient machine that automatically provides cash that employs us to perform the functions of a bank teller, and as partial employees, the bank sheds jobs as it cunningly, and automatically, charges us to work for it. The story of automation illustrates more artfulness than replacing bank tellers. From habitual behaviour to psychic writing to the spiritual in art to changing gears to self-loading weapons, even nature was once described as automated (plants and fruits growing without culture). To automate its theory, design must abandon rules, and needing no persuasion, and like nature, start to grow wild.

## Notes and Further Reading

Aaron Benanav, "Automation and the Future of Work—1", *New Left Review*, 119 Sept./Oct. 2019.

Aaron Benanav, "Automation and the Future of Work—2", *New Left Review*, 120 Nov./Dec. 2019.

Franco 'Bifo' Berardi, "The Neuroplastic Dilemma: Consciousness and Evolution", *e-flux Journal*, No. 60, December 2014.

#21

# Re

Repairable

# #21
# A Repairable Theory of Design[ing]

*Craig Bremner & Paul Rodgers*

Design has always been in a rush to replace what is still viable with completely new practices and projects, and new products and services. While novelty is design's lifeline, it is routinely derivative. But if you look at the history and practice of repairing things, it is clear that repair has always borne the weight of design's failings. So, with mounting evidence of its failings (*e.g.* landfill and bewildering and extortionate services), it would be far more accurate to consider design as something that doesn't need innovation; it needs constant repair. By compensating for the ongoing failure of design's current practices, repair represents hope, and by giving hope to what design has done in the past, it offsets moralistic reproval for the new. While repair balances the constant technological push for novelty, a repairable theory of design must recognise the fragility of design.

### Notes and Further Reading

Stephen Graham and Nigel Thrift, "Out of Order: Understanding Maintenance and Repair", *Theory, Culture and Society*, 24, (3), 2007, pp. 1-25.

Steven J. Jackson, "Rethinking Repair", In *Media Technologies: Essays on Communication, Materiality and Society*, edited by T. Gillespie, P. Boczkowski and K. Foot, Cambridge MA: The MIT Press, 2014.

#22

# Bu

Business

# #22
# A Business Theory of Design[ing]

*Paul Rodgers & Craig Bremner*

Design is big business. Although these two cultures are often very different, they are not irreconcilable. Design celebrates individuality and the credence given to creativity and business similarly accept these concepts as important. Success is an essentially conquering value, which implies the creation of wealth. Recognition and appreciation are also part of success, and designers generally acknowledge these values. In commercial realms, however, the notion of individuality and creativity are of little importance - what matters is profit. Surprisingly, however, the etymology of the word 'business' does not equate to capitalism. Business has its roots in the late fourteenth-century word *bisignes*, referring to "care, anxiety, occupation," and *bisig*, "careful, anxious, busy, occupied, diligent." Yet its Anglo-Saxon roots, rather than its Latin ones, are significant in helping to signal a pejorative ethic of "work, occupation" leading to the introduction of busi-ness - not lazi-ness. It's the claim to be busy that has perversely become a dominant virtue within capitalism (as well as in design). "*Sorry, I'm busy...*" has become the religious mantra of the worker - but also the designer and the cultural producer - as Franco Berardi (2011) articulates in his "Time, Acceleration, and Violence" essay.

### Notes and Further Reading

Franco (Bifo) Berardi, "Time, Acceleration, and Violence", *e-flux Journal*, No. 27, September 2011, available at: http://www.e-flux.com/journal/time-acceleration-and-violence/

Luc Boltanski and Arnaud Esquerre, "The Economic Life of Things: Commodities, Collectibles, Assets", *New Left Review*, Vol. 98, March 2016, pp. 31 - 54.

David Graeber, *Debt: The First 5,000 Years*, New York, NY: Melville House, 2011.

#23

# Re

Responsive

# #23
# A Responsive Theory of Design[ing]

*Craig Bremner & Paul Rodgers*

In response to the terminal condition of design, Dieter Rams and Ettore Sottsass tried to remind us - perhaps too long ago for most memories - that design is a serious business (Rams, 2011). Finally, their message seems to be penetrating – or is it? We learn that LEGO are into Serious Play… but to build better businesses. Design will style the leadership of these seriously better businesses to be creative. Then with seriously creative leaders at the helm of the Capital odyssey East will be, by design, branded West (or vice versa – design can't be sure) and unified – a world first. This is true because, through the design of data visualisation glasses, we will finally get to observe the real world. Which is only a problem for the design of healthcare because it claims to be looking through patient's eyes. Clearly, a responsive theory of design will have some challenges.

## Notes and Further Reading

https://www.lego.com/en-us/seriousplay

Matt Warman, "Dieter Rams: Apple has Achieved Something I Never Did," *The Telegraph*, June 7, 2011, www.telegraph.co.uk/technology/apple/8555503/Dieter-Rams-Apple-hasachieved-something-I-never-did.html (accessed March 13, 2012).

#24

# Sh

Sharing

# #24
# A Sharing Theory of Design[ing]

*Craig Bremner & Paul Rodgers*

The sharing-economy is what sociologist Eric Olin Wright calls "... *capitalism between consenting adults*" (Wright, 2013: 7). Adults consenting to all forms of sex gave birth to the permissive society. Consenting to all forms of capitalism has given birth to the digital platforms of the gig-economy where people go into debt, sucking value from households, families, communities and nature, to work under progressively precarious conditions. By sustaining the platform of the digital economy, design is increasingly isolated from the threadbare fabric of the society it is supposed to serve. In its place, design serves the exponential growth of venture capital that under the guise of the sharing-economy corrodes employment and wages substituting it with precarious service work. A sharing theory of design will have to reconcile design's consenting indebtedness to capitalism with its rhetoric about caring for the one world we share.

## Notes and Further Reading

Peter Frase, "Four Futures: One thing we can be certain of is that capitalism will end", *Jacobin Magazine*, December 13 2011, https://www.jacobinmag.com/2011/12/four-futures/

Nancy Fraser, "Contradictions Of Capital And Care", *New Left Review*, 100 July/ August 2016.

Eric Olin Wright "Transforming Capitalism Through Real Utopias", *American Sociological Review,* 78, 1, 2013, pp. 7.

#25

# Cl

Climatic

# #25

# A Climatic Theory of Design[ing]

*Craig Bremner & Paul Rodgers*

The weather is an archive of elements in constant and infinite reassembly via complex variables, which are also part of the archive. The weather archive was established as the planet came to life and is more or less known - it is called the climate. Life on earth was part of the weather archive, but civilisation consuming the landscape, then industrialisation consuming the natural resources, and now financial capitalism consuming the future, were never part of the archive. This squandering has forged new variables affecting the climate - new variables that are creating new extremes in the weather that have created new virtual risk environments that are now traded as future weather derivatives. We now live under a climate designed by capital in an atmosphere changed by design. A climatic theory of design[ing] can only really hope to salvage a better forecast for the strange weather ahead.

## Notes and Further Reading

Andrew Blum, *The Weather Machine: A Journey Inside the Forecast*, New York, NY: HarperCollins, 2019.

Phu Hoang, "Designing for Future Weathers", *Places Journal*, (no date), Available at: https://placesjournal.org/reading-list/designing-for-future-weathers/?cn-reloaded=1

Samuel Randalls, "Weather Profits: Weather Derivatives and the Commercialization of Meteorology", *Social Studies of Science*, 40, (5), 2010, pp. 705–730.

Andrew Ross, *Strange Weather: Culture, Science and Technology in the Age of Limits*, London, UK: Verso Books, 1991.

#26

# Mi

Misfit

# #26

# A Misfit Theory of Design[ing]

*Paul Rodgers & Craig Bremner*

Every design problem begins with an effort to achieve fitness between two entities - the form and its context according to Christopher Alexander, the author of the seminal book Notes on the Synthesis of Form (Alexander, 1971). The form, which the designer has control over, may be the solution to a problem whereas the context, which the designer has no real control over, frames the problem. Alexander reminds us that when we speak of design, however, the real object of discussion is not the form alone, but the ensemble comprising the form and the context. And as a designer, the real objective is to achieve a harmonious ensemble. When an ensemble works well, it has "good fit". We strive to put the context and form (ensemble) into effortless contact or frictionless coexistence. But it can be difficult to notice "good fit". It is much easier to see the lack of it. Or misfit. As Alexander states - "*It is the departure from norms that stand out in our minds rather than the norm itself. Their wrongness is somehow more immediate than the rightness of less peculiar behaviour, and therefore more compelling.*" (Alexander, 1971: 22). Accordingly, a misfit theory of design embraces wrongness in all its guises.

## Notes and Further Reading

Christopher Alexander, *Notes on the Synthesis of Form*, Cambridge, MA: Harvard University Press, 1971.

Christopher Alexander, Sara Ishikawa, Murray Silverstein with Max Jacobson, Ingrid Fiksdahl-King, Shlomo Angel, *A Pattern Language: Towns, Buildings, Construction*, Oxford, UK: Oxford University Press, 1978.

#27

# An

Anonymous

# #27 An Anonymous Theory of Design[ing]

*Paul Rodgers & Craig Bremner*

When asked which of the objects he designed was his favourite, Italian design maestro, Achille Castiglioni responded - "*The object I'm proudest of? A switch for an electric lead I designed thirty years ago with my brother Pier Giacomo. It was produced in large numbers and bought for its formal qualities, but no one knows who made the design.*" (Castiglioni, 2001: 252). What Castiglioni was talking about can, in a positive sense, be defined as design anonymity and utilitarianism. Today, navigating through the multitude of design and lifestyle magazines, websites and festivals in circulation, anonymity nowadays seems to be the designer's greatest fear. While many contemporary designers' relationships with utilitarianism are, at best, ambiguous – we should reassess whether or not anonymity today is a designer's biggest ambition or their greatest phobia? It would be interesting to know how many of today's superstar designers would be happy to remain anonymous.

## Notes and Further Reading

Sam Bowker and Craig Bremner (Editors), "Anonymous: The Void in Visual Culture", *Fusion Journal*, 009, September 2016, Available at https://fusion-journal.com/issue/009-anonymous-the-void-in-visual-culture/

Achille Castiglioni, "The Object I'm Proudest of? A Switch for an Electric Lead I Designed 30 Years Ago", in Sergio Polano, *Achille Castiglioni: Complete Works*, Milan: Electa, 2001.

#28

# Co

Comfortable

# #28
# A Comfortable Theory of Design[ing]

*Paul Rodgers & Craig Bremner*

Today, we live in an entirely artificial world where we routinely struggle to avoid things that have been designed. Generally held assumptions suggest that this makes us more superficial and more materialistic. The main assumption being that our relationships with things come at the expense of our relationships with people. In Daniel Miller's beautiful book, "The Comfort of Things", he carefully points out, however, that the opposite tends to be true (Miller, 2009). Our possessions often remain profound, and usually, the closer our relationships are with objects, the closer our relationships are with people. It turns out that the things that matter most to people are material things – their photographs, their music collection, their china figures, and their furniture. Their relationships with these objects become central to their relationships with people – their partners, their children, their brothers and sisters, and their friends and workmates. As designers, we need to be mindful then that people make sense of their day-to-day lives through their collection of objects.

### Notes and Further Reading

Mihaly Csikszentmihalyi and Eugene Rochberg-Halton, *The Meaning of Things: Domestic Symbols and the Self*, Cambridge: Cambridge University Press, 1981.

Daniel Miller, *The Comfort of Things*, London: Polity Press, 2009.

Susan Stewart, *On Longing: Narratives of the Miniature, the Gigantic, the Souvenir, the Collection*, Durham, NC: Duke University Press Books, 1992.

#29

# Pa

Paradoxical

# #29
# A Paradoxical Theory of Design[ing]

*Paul Rodgers & Craig Bremner*

Design is full of paradoxes. We know that design in its many guises attempts to make things lighter, faster, better, smoother, less expensive, sexier, etc. We also accept that design's raison d'être is to make the world a better place. The paradox, however, is that in design realms we are attempting to sustain the unsustainable, disciplining the undisciplined, proposing future visions that are in effect very harsh realities. However, whilst design researchers have been probing various aspects of design, it is highly likely that they have been looking in the wrong places. A paradoxical theory of design posits another way of thinking about design where the emphasis is on understanding how little we know and recognising that the unknown is a necessary and welcome condition for the beginning of any design research pursuit. That is, as designers, we should focus on what we don't know (Hara, 2015) and move away from current forms of design that largely results in imitation and derivation.

## Notes and Further Reading

Kenya Hara, *Ex-formation*, Zurich: Lars Müller Publishers, 2015.

Paul A. Rodgers, Giovanni Innella and Craig Bremner, "Paradoxes in Design Thinking", *The Design Journal*, Vol. 20, Supp 1, S4444 - S4458, DOI: 10.1080/14606925.2017.1352941.

#30

# Ma

Matriarchal

# #30
# A Matriarchal Theory of Design[ing]

*Craig Bremner & Paul Rodgers*

Through its ingenuous weakness for change driven by technology, it must be clear that design is fixated on technological determinism. The spectre of techno-progress has distracted design from comprehending that while technological change delivers not only voguish 'disruption' it also distorts socioeconomic conditions, warps communal relations and robs planetary capacity. The fetish for 'apps' seems to be driven by "*20-something, largely male coding disruptors in Silicon Valley fashioning their ideas of a better future based on apps and on-demand digital services for things to do what their moms used to do—wash their clothes, drive to the store, clean their dishes, buy their food. They invent apps to do it for them but what they think is progress has been labelled as mom tech.*" (Cohen, 2017). Mom tech has disrupted the historic gender bias in the DNA of design, giving birth to a matriarchal theory of design.

### Notes and Further Reading

Benjamin R. Cohen, "Jill Lepore On the Challenge Of Explaining Things", Public Books, 24 April 2017, http://www.publicbooks.org/jill-lepore- on-the-challenge-of-explaining-things/

Rebecca Solnit, *Men Explain Things To Me*, Chicago, IL: Haymarket Books, 2014.

#31

# Da

Dangerous

# #31
# A Dangerous Theory of Design[ing]

*Craig Bremner & Paul Rodgers*

In Richard Redgrave's writings compiled by his brother and published in 1876 as a Manual of Design, he warned that the allure of rich materials "*...give a meretricious splendour viciously alluring and dangerous to young designers.*" (Redgrave, 1876: 21). In stark contrast, just over a century later Ezio Manzini was celebrating our capacity to invent materials. Redgrave was clearly concerned about how readily design could be seduced, but by the time Manzini was writing design was the celebrated seducer. The meretricious became conspicuous in the profusion of imitation; splendour announced the advent of the society of the spectacle; and danger was writ large in the desperation of young designers to scoop some visibility. But the 'viciously alluring' temptation for design is now aiding and abetting the design of violence, one of Capital's main branches of business. A theory of design that disregards this dangerous temptation is a crime.

## Notes and Further Reading

Ezio Manzini, *The Material of Invention: Materials and Design*, Rome, Italy: Arcadia Edizioni, 1986.

Richard Redgrave, *Manual of Design, Compiled from the Writings and Address of Richard Redgrave, by Gilbert R. Redgrave*, London, UK: Chapman and Hall, 1876.

Richard Redgrave and Gilbert Richard Redgrave, *Manual of Design, Compiled from the Writings and Address of Richard Redgrave*, London, UK: Chapman and Hall, 1876.

#32

# Je

Jealous

# #32
# A Jealous Theory of Design[ing]

*Paul Rodgers & Craig Bremner*

Jealousy has long been associated with design. In the ancient Greek tale of the jealous potter who is so keen to guard her competitive advantage that she never reveals knowledge of the throwing technique, glaze formulae or painting as this potential generosity would devalue the uniqueness and hence negate the authority of the potter's work (Nelson, 2009). In the commercial realm, the motives for protecting copyright are powerful and may override any sense of collegiality in an organisation. But jealousy is not all bad. Among the Greeks, the concept of jealousy was overwhelmingly positive, more closely approximating zeal than jealousy as understood today. In other contexts, the idea of jealousy equates to notions of pride, honour or glory. In Greek, there is also a verb to make oneself jealous over something which can also express esteem, admiration or praise and these conceptions extend to an abstract noun describing a creative act or work of art which is worthy of jealousy, the object of emulation, envy or ambition, high fortune, which today can definitely apply to a lot of contemporary design. So, the measure of good design perhaps is the degree of envy with which it raises in a competitor.

## Notes and Further Reading

Robert Nelson, *The Jealousy of Ideas: Research Methods in the Creative Arts*, Ellikon, Australia, 2009.

Ana Quintela Ribeiro Neves Ramalho, "Will Robots Rule the (Artistic) World? A Proposed Model for the Legal Status of Creations by Artificial Intelligence Systems, *Journal of Internet Law*, Vol. 21, Issue 1, 2017, pp. 12 - 25.

McKenzie Wark, Designs for a New World, *e-flux Journal*, No. 58, October 2014.

#33

# Mo

Mongrel

# #33
# A Mongrel Theory of Design[ing]

*Paul Rodgers & Craig Bremner*

Design just like Fine Art before it has undergone something of a major transformation in recent years. Design, too, has refocused its lens to privilege ideas over aesthetics. As such, today, design can be anything. Bruno Latour famously claimed that: "*...design has been expanding ferociously from the design of objects that we use on a daily basis to cities, landscapes, nations, cultures, bodies, genes, political systems, the way we produce food, to the way we travel, build cars and clone sheep.*" (Latour, 2008: 2-3). Moreover, if you study how design is celebrated nowadays by the likes of the UK's Design Council, then its 'winners' routinely range from things like drugs that enhance sexual performance (Viagra) to business software. Stuart MacDonald (2012) describes this new creative landscape as a "*post-modern soup*" in which cultural, economic, social and educational issues are swimming and where "*mongrel*" institutions will flourish. But if design can be anything, then it can also be nothing, and this perhaps is the biggest challenge that design now faces? As the title of Arthur Danto's essay (1998) goes (if we substitute the word 'art' for 'design'), "After the End of Design" what does it mean when design can be anything?

## Notes and Further Reading

Arthur C. Danto, *After the End of Art*, Princeton: Princeton University Press, 1998.

Bruno Latour, "A Cautious Prometheus? A Few Steps Towards a Philosophy of Design (With Special Attention to Peter Sloterdijk)", *Keynote Lecture for the Networks of Design meeting of the Design History Society*, Falmouth, Cornwall, 3rd September 2008.

Stuart MacDonald, *Designs on Democracy: Architecture and Design in Scotland Post Devolution*, England: Zero Books, 2012.

#34

# Ti

Tired

# #34
# A Tired Theory of Design[ing]

*Paul Rodgers & Craig Bremner*

Franco Berardi, aka "Bifo", founder of the renowned "Radio Alice" in Bologna and an important figure of the Italian Autonomia Movement points out that the notion of being tired has always been a bugbear to the dream of modernism, the endless thirst for economic growth and profit, and the denial of organic limits (Berardi, 2011). Berardi believes, however, that our future will not be driven by energy, but by slowness, withdrawal, and tiredness and he reminds us that we were first advised of the finite physical resources of the planet when the Club of Rome commissioned the book The Limits to Growth over 40 years ago in 1972 (Meadows *et al.*, 1972). We have a world that is seriously unprepared to deal with the mounting crises we face because we have based our ways of life on the identification of energy and good, have an overriding obsession with accumulation, property, and greed, and strive for continual expansion and social well-being. But if we were to contemplate a creative consciousness of tiredness, as Berardi proposes, the current crises may mark the beginning of a massive abandonment of competition, consumerist drive, and dependence on work and help address the contemporary malaise.

## Notes and Further Reading

Franco (Bifo) Berardi, "Time, Acceleration, and Violence", *e-flux Journal*, No. 27, September 2011, available at: http://www.e-flux.com/journal/time-acceleration-and-violence/

Dennis Meadows, Donella Meadows, Jørgen Randers, William W. Behrens III, *The Limits to Growth: A Report for the Club of Rome's Project on the Predicament of Mankind.* Universe Books, 1972.

#35

# Wi

Winning

# #35
# A Winning Theory of Design[ing]

*Craig Bremner & Paul Rodgers*

The rise of global sport brought winning into the living rooms of everyone on the planet with a television and television quickly realised there are almost infinite profits in winning as a genre. Already very familiar with game shows, television started to structure programs where contestants won on-going participation, immunity, and lifelines resulting first in winning opportunities then ultimately winning celebrity. A winner was the product of the failure of others, and we have an insatiable appetite for the failure of others. In terms of production costs, failure comes cheap. What television first called 'reality' TV morphed into 'structured reality' where, by design, the fake is substituted for the real. That it is so easy to substitute the fake for the real reveals that again, by design, everything can be faked and, in theory, any failure will still result in a win for design.

## Notes and Further Reading

Andy Miah, *Sport 2.0: Transforming Sports for a Digital World*, Cambridge, MA: The MIT Press, 2017.

Steve Redhead, *Football and Accelerated Culture: This Modern Sporting Life*, London, UK: Routledge, 2015.

Rory Smith, "Is Football Still Sport?", *The Blizzard*, 1st September 2011, Available from https://www.theblizzard.co.uk/article/football-still-sport

John Thackara, *Winners! How Today's Successful Companies Innovate by Design*, Aldershot, UK: Gower Publishing Ltd., 1997.

#36

# Zo

Zombie

# #36
# A Zombie Theory of Design[ing]

*Craig Bremner & Paul Rodgers*

The steady B-Grade film career of zombies mushroomed with the explosion of digital games. Killing zombies has proven to be hugely popular and hugely profitable, and design is very culpable in their death. Still, the zombies keep coming as do design graduates working from their bedrooms designing games and hoping, like a TV Apprentice, not to be eliminated from this entrepreneurial bonanza. Working for next to nothing while looking to be the next wave of innovative magnets of venture capital is a precarious existence fuelled by the dream of being the next digital design impresario. But this dream is not going to come true from designing entertaining ways to kill zombies because selling one's skill for very cheap leaves design resembling a soulless corpse – which is the definition of a zombie – producing a very jaundiced theory of design.

## Notes and Further Reading

Benjamin Bratton, "Some Trace Effects of the Post-Anthropocene: On Accelerationist Geopolitical Aesthetics", *e-flux Journal*, No. 46, June 2013.

Timothy Morton, *Hyperorbjects: Philosophy and Ecology after the End of the World*, Minneapolis, MN: University Of Minnesota Press, 2013.

John Quiggin, *Zombie Economics: How Dead Ideas Still Walk Among Us*, Princeton, NJ: Princeton University Press, 2010.

Gerald Raunig, Gene Ray and Ulf Wuggenig (Editors), *Critique of Creativity Precarity, Subjectivity and Resistance in the 'Creative Industries'*, London, UK: MayFlyBooks, 2011.

Suely Rolnik, "Zombie Anthropophagy", in I. Curlin and N. Ilic (Editors), *Collective Creativity*, Kassel, Germany: Kunsthalle Fridericianum, 2005.

#37

# Sw

Swatchification

# #37
# A Swatchification Theory of Design[ing]

*Paul Rodgers & Craig Bremner*

Watches used to be a once in a lifetime purchase. In fact, often, watches were considered an heirloom that would be passed from one generation to the next. In the late 1970s, however, the world became flooded with watches from Asia equipped with quartz movements that kept good time and were as accurate as the best Swiss mechanical watches. More importantly, they were cheap. In a few short years, the value of Swiss watch exports was cut in half. The Swiss share of the watch market dropped from over 50 to 15 per cent, and competition from Asia slashed the number of watchmaking jobs in Switzerland from 90,000 to fewer than 25,000. In 1983, Nicolas G. Hayek turned the world of watch manufacture upside down with his innovative and radical proposal of a 'second watch'. The idea was not to produce an expensive piece of well-crafted jewellery, but a new inexpensive and innovative watch that would be provocative, fashionable, desirable, and didn't cost a fortune. In 2006, Swatch celebrated the production of its 333 millionth Swatch watch and continues to enjoy its position as one of the biggest brand names in the world, known everywhere as a maker of entertaining and exciting products that are in tune with the times. Perhaps what the world needs now is the 'Swatchification' of how we plan and deliver our health, transport, political, environmental, and social services?

## Notes and Further Reading

Pierre-Yves Donzé, *A Business History of the Swatch Group: The Rebirth of Swiss Watchmaking and the Globalization of the Luxury Industry*, London, UK: Palgrave Macmillan, 2014.

Thomas Hauffe, *Design: A Concise History*, London, UK: Laurence King Publishing, 1998.

#38

# An

Anti-Design

# #38 An Anti-Design Theory of Design[ing]

*Paul Rodgers & Craig Bremner*

Anti-Design originated in Italy in the 1960s as a reaction against what many designers saw as the impoverished language of Modernism. Anti-Design was motivated particularly by the increasingly widespread dwindling of the social relevance of design at the expense of capitalist enterprise. Modernism's emphasis on style and the aesthetics of good form was anathema to the anti-design movement. Two key exponents of Anti-Design, Archigram and Superstudio, expressed their ideas in the production of furniture prototypes, exhibition pieces, and the publication of manifestos aiming to harness the social and cultural potential of design rather than embrace style and good form as a means of increasing sales. Anti-Design challenged Modernism's notions of permanence, muted colour palette, integrity of materials, and the motto 'form follows function' by embracing the ephemeral nature of Pop consumerism, the rich potential of colour, ornament and decoration, and the expressive potential of kitsch, irony, and distortion of scale respectively. A more recent view of anti-design has been promoted by Martyn Perks who argues that the greening of design seeks to deliberately curtail the freedom of designers to think and that all designers should strive to make a virtue out of doing less and thinking small. This is 'Anti-Design' proclaims Perks. Holding back inevitably means producing crap, and that affects us all. We know that design fundamentally involves analysing and synthesising ways of planning and doing things. Design challenges the status quo. It confronts convention and preconceived notions. And this involves challenging commonly held assumptions such as sustainable design.

## Notes and Further Reading

Martyn Perks, "The Rise and Rise of Anti-Design", *Spiked Online*, accessed 21 December 2018: https://www.spiked-online.com/2008/08/21/the-rise-and-rise-of-anti-design/

#39

# Fa

False

# #39
# A False Theory of Design[ing]

*Craig Bremner & Paul Rodgers*

Because so much of what used to be called news cascades through Facebook and people believe so much of it, and so much of this news is fabricated, eventually Facebook had to send to its people (formerly known as users) "10 Tips to Spot False News". Facebook people are warned to 'watch for unusual formatting' (tip number 4) and manipulated images ('consider the photos', tip number 5), which help make a story unbelievable ('be skeptical of headlines', tip number 1) and 'intentionally false' (tip number 10). Facebook's people need to think critically about the stories they read and ask themselves - is the story a joke? (tip number 9), because sometimes false news stories can be hard to distinguish from humour or satire. Sound advice. But did you spot that tips number 4 and 5 are typically done by design? Let's face it. Design is complicit in fabricating false news, meaning a false theory of design is no joke.

## Notes and Further Reading

Barbara Cassin, Emily Apter, Jacques Lezra, and Michael Wood (Editors), *Dictionary of Untranslatables: A Philosophical Lexicon*, Princeton, NJ: Princeton University Press, 2014.

https://www.facebook.com/help/188118808357379

Geert Lovink, "Overcoming Internet Disillusionment: On the Principles of Meme Design", *e-flux Journal*, No. 83, June 2017.

#40

# Se

Sentimental

# #40
# A Sentimental Theory of Design[ing]

*Craig Bremner & Paul Rodgers*

Experiences are filtered through senses whose richness and complexity we describe to others using the language of feelings or sentiment. Once flattering, now indulgent, sentiment has become a superficial emotion. Sentiment spawns sentimentality and being sentimental is misrepresenting the world in order to indulge our feelings. But Jacques Ranciere reminds us that in fact, the language of sentiment preceded that of reason so when in the 1980s Alessandro Mendini proposed designers become "sentimental robots", he implored design to stop indulging the rational competitive frenzy and wilfully misrepresent and reject design as it filters through our senses. Now we are warned, robots are exiting the factory to take over the world. It is estimated that over the next two decades, between 45 and 57 percent of jobs in the OECD could be automated. If that happens, design will no longer need theories. It will become a sentimental practice.

## Notes and Further Reading

Alessandro Mendini, "Caro Giovanni Designer", *Domus*, 648, 1984.

Mary Midgley, "Brutality and Sentimentality", *Philosophy*, Vol. 54, No. 209, July 1979.

Ljubica Nedelkoska and Glenda Quintini, "Automation, Skills Use and Training", OECD Library, DELSA/ELSA/WD/SEM(2018).

Jacques Ranciere, *The Ignorant Schoolmaster: Five Lessons in Intellectual Emancipation*, Stanford: Stanford University Press, 1991.

#41

# Fa

Fantastic

# #41
# A Fantastic Theory of Design[ing]

*Craig Bremner & Paul Rodgers*

A fantastic theory of design is closely related to the fantasy of design; both skirt the indefinable but from different directions. Like fantasy sport (where fantasy players perform for millions of people who wage billions to win an imaginary game), design wages its performance on an imaginary game of making things better. Overwhelmingly, the evidence exposes that design is a fantasy of talent, of creativity, of innovation, and of disruption luxuriating on a fantasy that design is fashioning a better world. In contrast, a fantastic theory of design might be indefinable but only by starting from the indefinite can design escape its fantasy reproduction of the world-as-found. The fantastic makes visible an imaginary but inescapable space where the world in which we live has to be designed in parallel with the worlds of yesterday and tomorrow. What fantastic prospects!

### Notes and Further Reading

Caroline Bassett, Ed Steinmueller & George Voss, *Better Made Up: The Mutual Influence of Science Fiction and Innovation*, Nesta Working Paper No. 13/07, March 2013, available at: https://media.nesta.org.uk/documents/better_made_up_the_mutual_influence_of_science_fiction_and_innovation.pdf

Jill Lepore, "The Disruption Machine: What the Gospel of Innovation gets Wrong", *The New Yorker*, 23 June, 2014, available at: https://www.newyorker.com/magazine/2014/06/23/the-disruption-machine

Paul A. Rodgers and Craig Bremner, "An A to Z of Contemporary Design", *The Design Journal*, Vol. 19, No. 1, 2016, pp. 5 - 23, DOI: 10.1080/14606925.2016.1109208

#42

# Pl

Plastic

# #42
# A Plastic Theory of Design[ing]

*Craig Bremner & Paul Rodgers*

When Mondrian proclaimed Neoplasticism his pure abstraction simultaneously liberated design from its sentimental adherence to naturalness and enslaved products to pure appearance. Then plastic and design embarked on a long, carefree and productive attachment during which plastic was heralded as the "*material of invention*" (Manzini, 1989). On the back of what design could do with plastic several developing nations remodelled themselves into industrial power-houses filling up our homes then landfill and now oceans (there is enough plastic in the oceans to cover an area the size of Russia). Plastic made design look creative, but like plastic, design is impressionable and easily manipulated, resulting in the plastic-coated redesign of purely superficial things. A plastic theory of design would have to remove the plastic-coating from design to restore its plastic merit: a plastic design that would exhibit the ability to adapt to planetary change and in the process learn.

## Notes and Further Reading

Ezio Manzini, *The Material of Invention: Materials and Design*, Cambridge, MA: The MIT Press, 1989.

Piet Mondrian's Neoplasticism in Painting (1917-1918), *In* Hans Ludwig C. Jaffé, *De Stijl*, New York, NY: H.N. Abrams, 1971.

#43

# Gr

Green

# #43
# A Green Theory of Design[ing]

*Craig Bremner & Paul Rodgers*

A theory of green design runs into semantic problems immediately. Being both noun and verb the tricks design has up its sleeve are well-rehearsed. As a noun design is being given credit for appearing to be greenish (*e.g.* we believe that green qualifies design's support for the protection of the environment). As a verb, it is green that is being designed (*e.g.* into an environmental movement). But green can also qualify the appearance of design (*i.e.* design's apparent greenness / immaturity / inexperience). What about green, which can be noun, verb and adjective? What happens to green design if looked at from the point of view of green? If green is the verb in green design, then it casts design in an unmistakable green or immature or inexperienced humour. This organic naivety is also unmistakable when design's conspicuous materialistic practice is juxtaposed with its shameless, holistic temper.

## Notes and Further Reading

Jeremy Davies, "Sustainable Nostalgia", *Memory Studies*, Vol. 3, Issue 3, 2010, pp. 262 – 268.

Dorothy Mackenzie, *Green Design: Design for the Environment*, London, UK: Laurence King Publishing, 1997 (2nd Revised edition).

Victor Papanek, *Design for the Real World: Human Ecology and Social Change*, London, UK: Thames and Hudson, 1985 (2nd edition).

Raj Patel & Jason W. Moore, *A History of the World in Seven Cheap Things: A Guide to Capitalism, Nature, and the Future of the Planet*, Berkeley, CA: University of California Press, 2020.

Pope Francis, *Laudato Si': On Care for Our Common Home (Encyclical)*, 24 May 2015, available at http://w2.vatican.va/content/francesco/en/encyclicals/documents/papa-francesco_20150524_enciclica-laudato-si.html

#44

# Da

Dark

# #44
# A Dark Theory of Design[ing]

*Craig Bremner & Paul Rodgers*

An already dark world is becoming darker evident in darkening disciplines such as dark ecology and dark anthropology. The invisible hand operating the market of economics has always lent a helping hand to its dark side – the 'black market'. The unregulated flow of 'dark money' is designed primarily to exploit democracy. Facebook has designed 'dark posts' as test beds for commerce that it promotes as 'unpublished'. The hidden Internet can be seen through the 'dark web'. 'Dark energy' and 'dark matter' fill big holes in the cosmic imaginary. In contrast, 'dark science' and 'dark theory' pick holes in science. Long before these darkenings, Vance Packard's "The Hidden Persuaders" (first published in 1957) exposed the dark patterns employed in the dark art of design. If a dark theory of design were to peer into how design has kept itself in the dark, then it might reveal something about the ignorance of design.

## Notes and Further Reading

Ranulph Glanville, "Inside Every White Box There are Two Black Boxes Trying to Get Out", *Behavioural Science*, Volume 27, Issue 1, January 1982, pp. 1 - 11.

Vance Packard, *The Hidden Persuaders*, New York, NY: Ig Publishing, 1957.

#45

# Gr

Grey

# #45
# A Grey Theory of Design[ing]

*Craig Bremner & Paul Rodgers*

When Alessandro Mendini wrote about "Black Design", he asked us to see the black inside ourselves. His radical design comrade, Andrea Branzi, censured the pursuit of standardisation as the "white whale" of design. Inwardly black but ordinarily white, when blended they render design grey. In a similar vein, Paul Virilio wrote "Grey Ecology" (2010) to shake our faith in green ecology. Green ecology tackles the pollution of the natural environment on which all living things depend. Grey ecology chronicles the pollution of the designed environment that humans cannot live without. In this artificial ecology, a grey theory of design might help retrieve design from its obsession with depicting its actions in black and white – problems lead to solutions, creativity leads to innovation, media leads to fame and fortune. Instead, perhaps a grey theory of design might intensify the blurry designerliness of design.

## Notes and Further Reading

Alessandro Mendini, "Caro Giovanni Designer", *Domus*, 648, 1984.

Paul Virilio, *Grey Ecology*, New York, NY: Atropos Press, 2010.

Julia Carrie Wong, "Greyball: How Uber used Secret Software to Dodge the Law", *The Guardian*, 4 March 2017, https://www.theguardian.com/technology/2017/mar/03/uber-secret-program-greyball-resignation- ed-baker

#46

# Fa

FabLab

# #46
# A FabLab Theory of Design[ing]

*Paul Rodgers & Craig Bremner*

We are constantly reminded of the ease nowadays of turning our future visions into real products. Personal fabrication systems, hackerspaces, makerspaces, FabLabs and their commercial counterparts like TechShops, co-working spaces, accelerators and incubators, it is claimed, give us all the ability to design and produce our own products from the comfort of our very own homes, which will revolutionize the world (Gershenfeld, 2007). Despite the widespread marketing clangour of the FabLab movement, these shared hacker and makerspaces can only be described accurately as "fringe phenomena" since they play a largely minor role in the production of wealth, knowledge, political consensus and the social organisation of life. In Issue 5 of the Journal of Peer Production (October, 2014), a stinging critique states that: "*Shared Machine Shops are not new...; Fab Labs are not about technology...; Sharing is not happening...; Hackerspaces are not open...; Technology is not neutral...; Hackerspaces are not solving problems...; and FabLabs are not the seeds of a revolution...*". (Maxigas and Troxler, 2014). The paradox here is that the easier it may be to design stuff today, the harder it is to design stuff well. Perhaps we should heed the warning from Dieter Rams here and take greater care with design and remember that design is far from easy.

### Notes and Further Reading

Neil Gershenfeld, *Fab: The Coming Revolution on Your Desktop - From Personal Computers to Personal Fabrication*, New York, NY: Basic Books, 2007.

Neil Gershenfeld, Alan Gershenfeld and Joel Cutcher-Gershenfeld, *Designing Reality: How to Survive and Thrive in the Third Digital Revolution*, New York, NY: Basic Books, 2017.

Maxigas and Peter Troxler, P. (2014). "Editorial Note: We Now have the Means of Production, but Where is my Revolution?", *Journal of Peer Production*, Issue 5, October 2014, http://peerproduction.net/ issues/issue-5-shared-machine-shops/

#47

# Ba

Barbaric

# #47
# A Barbaric Theory of Design[ing]

*Craig Bremner & Paul Rodgers*

It is more than evident that the task of design has not been concerned with producing a better world. It has been concerned with alleviating anxiety about consuming the world. So, when authorities in New Orleans gave shelter to the rich and abandoned the poor to Hurricane Katrina (a category 5 hurricane that made landfall on Florida and Louisiana in August 2005), it illustrated how the riotously unequal consumption of the world is producing a new kind of barbarism. And in the endless stream of catastrophic futures predicted to plague the planet, many point to this coming barbarism, which will inevitably infect design. But to be a designer in a pandemic of inequality does not mean design has to be barbaric. However, a barbaric theory of design will be indispensable if only to counteract disaster capitalism's capacity to profit from the paradoxical, the contradictory, the accidental, and the future of the world – all once the remit of design.

## Notes and Further Reading

Isabelle Stengers, *In Catastrophic Times: Resisting the Coming Barbarism*, London, UK: Open Humanities Press, 2015.

Wen Stephenson, "Against Climate Barbarism: A Conversation with Naomi Klein", *Los Angeles Review of Books*, September 30, 2019, Available at: https://www.lareviewofbooks.org/article/against-climate-barbarism-a-conversation-with-naomi-klein

#48

# Ex

Exaggerated

# #48
# An Exaggerated Theory of Design[ing]

*Craig Bremner & Paul Rodgers*

Andrea Branzi concludes an interview about his autobiography "Una Generazione Esagerata" saying: "*...whoever doesn't exaggerate today is an insignificant minority within design.*" Exaggeration has replaced normality, which, says Branzi: "*...no longer has a value...and...in design today is impractical.*" (Campanini, 2014). In design terms, exaggeration describes a form of enhancement mostly employed first to confuse, then inflate the value of experiences, and when it comes to experience augmented reality is touted to be the climax. Quaintly described as the technique of adding digital graphics to a user's view of the physical world, augmented reality is yet another exaggerated design to enhance commerce. If a theory is derived from a spectator, then normally their observations would explain something. With normal having no value, an exaggerated theory of design is the perfect explanation of the military-entertainment complex whose theatre of operation is the mercantile spectacular.

## Notes and Further Reading

Andrea Branzi, *Una Generazione Esagerata*, Milano: Baldini & Castoldi, 2014.

Cristiana Campanini, "Branzi teorico del radical design: 'L'esagerazione esalta Milano'", *La Repubblica Milano*, 30 April 2014, http://milano.repubblica.it/cronaca/2014/04/30/news/branzi_teorico_del_radical_design_l_esagerazione_esalta_milano-84891223/

#49

# Po

Pornographic

# #49
# A Pornographic Theory of Design[ing]

*Craig Bremner & Paul Rodgers*

In a world "*...dominated by the merchandise of promises that are never fulfilled and always postponed... pleasure is advertised and promised, and simultaneously postponed and deceived.*" According to Franco 'Bifo' Berardi: "*...this is the pornographic feature of semioproduction in the sphere of the market.*" (Berardi, 2012: 4). It is no stretch of the imagination to see design actively exciting the mercantile spectacular to the stage where the production of 'signs' massively exceeds their ability to be consumed. In the pursuit of pleasure, we swim in the erotically designed semiosphere surrounded by impossible offers. The promise is an image designed for pleasure while in actuality, it is an impossible projection frustrating desire and climaxing in panic, depression and hate. If the word pornography is derived from the Greek for writing about prostitutes, then a pornographic theory of design must begin by asking 'who now is the prostitute?'.

## Notes and Further Reading

Franco "Bifo" Berardi, "Emancipation of the Sign: Poetry and Finance During the Twentieth Century", *e-flux Journal*, No. 39, November 2012, https://www.e-flux.com/journal/39/60284/emancipation-of-the-sign-poetry-and-finance-during-the-twentieth-century/

Thomas Meaney, "A Celebrity Philosopher Explains the Populist Insurgency", *The New Yorker*, February 26, 2018, https://www.newyorker.com/magazine/2018/02/26/a-celebrity-philosopher-explains-the- populist-insurgency

#50

# Cu

Curious

# #50
# A Curious Theory of Design[ing]

*Constantin Boym*

This theory believes in curiosity as the basis of design action. Designs of many generations – from children's books hero Curious George (1941) to Curiosity, the NASA rover on Mars (2015) paid homage to this endearing human trait. A curious theory of design[ing] is based on one's ability to see; to detect the intangible vibes of cultural sentiments; to observe people's habits and know-how; to discern beauty in the detritus of daily life. This theory calls for travel to far-away lands, appreciates activities of an urban flaneur, and acknowledges cultural shock as a productive phenomenon. Curious designers' working tools are a notebook, a camera, and an iPhone. Curious designers tinker and play with everything they find; they cross-pollinate, reuse, and rethink. A curious theory of design[ing] favours experimentation and embraces failure. Admitting that our world is a troubled place, this theory believes that even serious and controversial problems of today are better addressed with a sense of humour.

## Notes and Further Reading

Andrew Blauvelt (Editor), *Strangely Familiar: Design and Everyday Life*, Minneapolis, MN: Walker Art Centre, 2003.

Margret Rey and H.A. Rey, *Curious George*, Boston, MA: Houghton Mifflin, 1941.

#51

# Le

Learning

# #51
# A Learning Theory of Design[ing]

*Maya Dvash*

Since design is a field that is, by definition, concerned with the everyday and centred on user experiences, it is more accessible to the public than visual art, and has greater potential to initiate learning processes. Learning through design is a theory capable of expanding the museum's relevance to the community, as well as its generative power. A learning theory of design[ing] makes the museum relevant in the context of both aesthetic discourse and practice. In this theory, the museum operates on three complementary levels - (1) It presents knowledge, or data (not only objects on display, but also that produced by others invited to participate in museum activities); (2) It produces a critical discourse; (3) It provides a performative platform (the political potential of social involvement and production of new knowledge).

### Notes and Further Reading

Craig Bremner and Laura Bernadet, "The Museum of the Future: A Sedimentary Cloud", *The Design Journal*, 20, Supp. 1, 2017, pp. S3560-S3568.

Guy Debord, *Society of the Spectacle*, New York, NY: Zone Books, 1994.

Hal Foster, "The Archive without Museums", *October*, 77, Summer 1996, pp.97-119.

Robert Venturi, Denise Scott Brown and Steven Izenour, *Learning from Las Vegas*, Cambridge, MA: The MIT Press, 1972.

#52

# Th

Theoretical

# #52
# A Theoretical Theory of Design[ing]

*Chiara Alessi*

The Italian school system, thanks to its long tradition, has supported the distinction between practical subjects and theoretical subjects, including design courses and courses of theory, between doing and thinking, between oral tests and exercises. Thus, a history of design, according to the directions of the ministry, should have as its objective the knowledge of books, facts, names and places, while a project teacher should direct students to practice with design. Periodically, this pluri-centennial scaffolding has been undermined now by pretensions of empiricization, reducing everything to experiences of doing and finding, now - typically this is witnessed today - to currents of intellectualization that have made the search for the main theme of design. Alessandro Mendini once told me that as a boy, his teacher sent him to draw the facades of the Churches, of which he had to minutely reproduce not only the architectural elements but also the ornamental ones. Was that a theory or a practice lesson?

**Notes and Further Reading**

Chiara Alessi, *Dopo gli anni Zero. Il nuovo design Italiano*, Milan, Italy: Editori Laterza, 2014.

Chiara Alessi, *Design senza designer*, Milan, Italy: Editori Laterza, 2016.

Chiara Alessi, *Le caffettiere dei miei bisnonni. La fine delle icone nel design Italiano*, Torino, Italy: UTET, 2018.

Loredana Parmesani (Editor), *Alessandro Mendini: Scritti di Domenica*, Milan, Italy: Postmedia Books, 2016.

#53

# Pr

Practical

# #53
# A Practical Theory of Design[ing]

*Chiara Alessi*

The idea is that the sum of theoretical and practical skills, at the end of the schooling path, define a complete designer, with solid theoretical foundations and learned practical skills, that on one side there are the militants of design, whose task is to pass concrete information and on the other hand the design speculators, who must impart theoretical notions. But we know well - it would be enough to observe the real "method" not given to children but elected by children - that the question of "sum" is never the topic, that the question of learning is always a knowledge of knowledge and body, of manual and cognitive skills, that you learn to do, knowing and knowing, doing. Theory in children is always based on a practice that they experience. It is a practical theory. In a certain sense it is an approach that reminds us of the gesture of art: children, like artists, go to the essence of the action of planning that lies in looking ahead and having foreshadowings. It is in their nature! In fact, they are the greatest revolutionaries.

### Notes and Further Reading

Andrea Branzi, "We Are the Primitives", *Design Issues*, Vol. 3, No. 1 (Spring, 1986), pp. 23 - 27.

Tom Holert, "Hidden Labor and the Delight of Otherness: Design and Post-Capitalist Politics", *e-flux Journal*, No. 17, June-August 2010.

Marco Petroni and Giovanni Innella, *Going Real: The Value of Design in the Era of PostCapitalism*, Malaga, Spain: Vernon Press, 2019.

#54

# Ag

Against

# #54
# Against a Theory of Design[ing]

*Chiara Alessi*

A designer friend of mine confessed to me that he had been persecuted by this dichotomy, between doing well and thinking well, since he was a child, and for Christmas, his school was organizing a competition for the design of the best Nativity scene. Every year, with his carpenter dad, my friend created an endless and exacting miniaturization of the Nativity in interminable days of work, but he was punctually beaten by a companion who at last resolved the work required with a work of denunciation, the result of an idea flash and, wanting to keep to the delivery date, even "off topic" that was more incisive and effective, even if made in a more subtle way. But by translating the characters of the crib into containers of political, civic, and topical messages, in the end, the interpretation of the other child always won. My friend applied a solid platonic knowledge of design to practice, while the latter adhered to a practice of a theory of his own. Choosing which one more convinces us can tell us a lot about our idea of design.

### Notes and Further Reading

Geert Lovink, *Sad by Design: On Platform Nihilism*, London, UK: Pluto Press, 2019.

Emanuele Quinz, Contro l'oggetto: Conversazioni sul Design, Milano, Italy: Quodlibet, 2020.

Paul A. Rodgers and Craig Bremner, "The Design of Nothing: A Working Philosophy", in *Advancements in the Philosophy of Design*, edited by Pieter Vermaas and Stephane Vial, Berlin, Germany: Springer Press, 2018, pp. 549 – 564.

#55

# Da

Data

# #55
# A Data Theory of Design[ing]

*Will Holman*

Increasingly, 'design' is merely a shell for a database. The streaming service app's smooth UX is meaningless without the servers full of music and movies switching noiselessly in the background. The autonomous car, curvaceous and bold, is an expensive lawn ornament without decades of mapping data and complex software. Our phones, painstakingly milled from billets of aluminium, are expensive calculators without their invisible tethers to the internet. This was not always the case; a radio is a radio, a book is a book, a bike is a bike, a chair is a chair. The "intelligence" of our everyday objects thus comes at the price of incomprehensibility by the user and an inability to fix them when they are broken. This makes the world an unfamiliar, even frightening place; to remove these uncertainties, the data must become the design and become intelligible to the user.

## Notes and Further Reading

Ruha Benjamin, *Race After Technology: Abolitionist Tools for the New Jim Code*, New York, NY: John Wiley & Sons, 2019.

Kris De Decker, *Low-tech Magazine 2012-2018 (Volume 2 of Low-tech Magazine)*, Barcelona, Spain: Kris De Decker, 2019.

Jaime Snyder, *Operating Manual for Spaceship Earth*, Zurich: Lars Müller Publishers, 2008.

#56

# Ju

Justified

# #56
# A Justified Theory of Design[ing]

*Clive Dilnot*

Design is justified in its own mind, rarely in that of others. What explains this gap? It is this. Acted, design intuits what it – and to a degree, it alone – enacts in the process of designing. But from the "spectators" viewpoint, which sees only what is realized, not only is process invisible, the outcome is in any case trivial. Pleasant or ugly by turns, the contingency, arbitrariness, of what is realized convinces no-one. Only in exception does something happen that transcends essential scepticism. The misplaced confidence of one side and the dismissal of the other have a common root: the inability to grasp that relations and not essences are at the core of the dependency of the human manner of being. Subjectivity is null without its objective moment: 'objectivity,' whether thought materially or immaterially has no meaning save in relation to subjective life. Mediation, the negotiation of incommensurabilities, the 'search for good designs,' are the constituent requirement of modes of dependency on artifice. Design is justified, not when it blindly reproduces or enacts these conditions but when it brings them to light, permits a life that fails to know or to value its dependency, to grasp, even momentarily, the possibilities of being within, and not in spite of, the artificial.

### Notes and Further Reading

Giorgio Agamben, *The Coming Community*, Minnesota, USA: University Of Minnesota Press, 1993.

Giorgio Agamben, *The Man without Content*, Stanford, USA: Stanford University Press, 1999 (Chapters 8 & 9).

Tony Fry, Clive Dilnot and Susan C. Stewart, *Design and the Question of History*, London, UK: Bloomsbury Academic, 2015.

#57

# Ma

Mature

# #57
# A Mature Theory of Design[ing]

*Clive Dilnot*

Maturity, in the social-scientific sense, used to be defined by 'autonomy' (Kant). Today, we understand a reversal of this notion. Maturity for us is the necessary acceptance of dependency. Autonomy (subjective autonomy) was always an illusion. It's not-so-hidden dependence on the 'struggle for the rough and material things' ('without which there can be nothing fine and spiritual') always made it a minority interest. Those denied participation naturally rebelled against the very idea. Yesterday, this took the form as fascism. Today, it manifests as the demand for immaturity in all things – a demand that Silicon Valley no less than Donald Trump are happy to service. Design too often seems to serve this end. Yet there is a contradiction. Technology, like Trump, is endemically immature. This is why it is so destructive. Design, however, thought as designing, as process, contests immaturity. It does so in three ways. First, in negotiating (mediating) incommensurability design accepts the impossibility of the singular over-simple answer (for design, all problems are wicked). Second, that it is a 'conversation with the situation' (and today with the actors involved in that situation) is more than a cliché. Inherently committed to 'ways of getting things into better relationships with each other' design operates in 'two directions.' It refuses the dominance of one over the other. Third, in accepting dependency (on what we make) design contests (or could contest) the immaturity by which technology is deployed but not truly incorporated; by which it is insufficiently 'developed to cope with the elemental forces of society' (Benjamin, 2008).

## Notes and Further Reading

Walter Benjamin, *The Work of Art in the Age of Mechanical Reproduction*, London: Penguin (Great Ideas), 2008.

Susan Meld Shell, *Kant and the Limits of Autonomy*, Cambridge, MA: Harvard University Press, 2009.

#58

# Po

Political

# #58
# A Political Theory of Design[ing]

*Clive Dilnot*

The essence of politics is simple. What matters are the forms in which human material needs are met and controlled. Both the variety of needs and the forms through which they can be met are legion. But they are not, as a social historian (Stephen Yeo) has noted, the same ways. The National Health Service (UK), Obama Care, and a wholly deregulated, privatized system are all ways of meeting the need for health care. But their consequences and effects – for persons - differ, massively. 'A society in which one was dominant, and the other recessive, would not be the same as one in which the roles were reversed.' Politics today is largely a fight about such consequences. Crudely, it comes down to a fight between (privatized) profit and the interests of persons. Design enters this equation in two ways. First, because every (political) form is in some manner "designed," is shaped in relation to interests. Second, because all such forms comprise situations and hence an ethics, as well as a politics, is entailed here. ('To be faithful to the situation means: to treat it right to the limit of the possible. Or, if you prefer: to draw from the situation, to the greatest possible extent, the affirmative humanity that it contains'). Today, design is at the conjunction of these two moments. Its brief is to explore, from the standpoint of all, the possibilities of situations. In so doing it begins to contribute to the creation of a societies in which accumulation is made secondary to an economy of needs. Those who dislike this idea must be reminded that unless the latter can come to pass their expectation for seeing out this century in comfort is unlikely to come to pass.

### Notes and Further Reading

Tony Fry, *Design as Politics*, London, UK: Berg, 2010.

Stephen Yeo, *A New Life, the Religion of Socialism in Britain, 1883-1896: Alternatives to State Socialism*, Brighton, UK: Edward Everett Root, 2019.

#59

# Ci

Circumstantial

# #59 A Circumstantial Theory of Design[ing]

*Clive Dilnot*

Specified requirements; objective necessities; defined and presupposed constraints are acknowledged as determinants of a design. By contrast, "circumstances" seem weak: at once contingent, malleable, in the end subjective. Yet an observation might refute this too easy observation. Speaking about Picasso in 1911, the art historian Michael Baxandall writes at one point: "*The painter registers his individuality... by his particular perception of the circumstances he must address.*" Indeed, if one is to think of a painter "*expressing himself*", it is most of all here, in the analysis of his environment which schematically speaking precedes the process of painting itself, that one can most securely locate an individuality. (Baxandall, 1987: 47). There is often a curious impersonality about the actual working out of a solution in the medium. ... But the painter's formulation of a brief is a very personal affair indeed. What applies to the artist applies even more to the designer. Design is always in truth an apperception of circumstances and the possibilities and limitations they offer. It is the acuity of the perception of circumstance (and what they permit) that grounds all originality in design. It is not that the perception of possibility negates objective necessities and constraints. It is that grasping what circumstances require and makes possible allows in Simon's language, the real "*scarce factors*" and "*limiting resources*" – and thus the real plenitude and possibility in the situation to be grasped. Designers who grasp circumstance open themselves to origination. All exemplary work in design follows this law.

## Notes and Further Reading

Michael Baxandall, *Patterns of Intention: On the Historical Explanation of Pictures*, New Haven, Connecticut: Yale University Press, 1987.

Herbert A. Simon, *The Sciences of the Artificial*, Cambridge, MA: The MIT Press, 1981.

#60

# Po

Power

# #60
# A Power Theory of Design[ing]

*Marco Petroni*

There is today wide agreement that design requires a reformulation in terms of an extension and deepening of its attitudes. A theoretical strategy needs to carry out such a task. There are some keywords that put design into jeopardy. Power is one of these. My objective is to establish a dialogue between these theoretical developments and politics. I believe that an anti-essentialist theoretical stand is the *sine qua non* (an indispensable and essential action) of a new vision for the design conceived in terms of a radical and plural field. The sense of the change, however, is difficult to determine. Without ultimate meanings, we can work on an expanded field as a consciousness of the contingent, precarious, limited character of power in design.

## Notes and Further Reading

Ulrich Beck, *Power in the Global Age: A New Global Political Economy*, Cambridge, MA: Polity Press, 2005.

Donna J. Haraway, *Staying with the Trouble: Making Kin in the Chthulucene*, London, UK: Duke University Press, 2016.

Graham Harman, *Object-Oriented Ontology: A New Theory of Everything*, London, UK: Penguin Books, 2018.

Graham Harman, *Art and Objects*, Cambridge UK: Polity Press, 2019.

#61

# Ed

Educational

# #61 An Educational Theory of Design[ing]

*Marco Petroni*

Everyone with a mobile phone reveals their location - whether or not the phone is equipped with explicitly locative technology, whether or not the phone is even turned on. Are the books and our education, the ways to know? Any information or potential insight bound up in the flow of events falls to the ground like a silent, diffuse drizzle, forever lost to introspection, analysis, and memory. Personally, I am at war with my time, with history, with all authority that resides in fixed and frightened forms. And this is a precarious condition. It is a way to install a new point of view of education. It is not simply the condition of millions of people who work without a permanent contract or without any guarantee of what will happen to their salary the following day. It is the condition in which every night is the last night, awakening from which each one of us sinks into the carriage of an underground train in order to perform their solitary, hyper-connected journey.

## Notes and Further Reading

John R. Blakinger, *Gyorgy Kepes: Undreaming the Bauhaus*, Cambridge, MA: The MIT Press, 2019.

Donna J. Haraway, *Staying with the Trouble: Making Kin in the Chthulucene*, London, UK: Duke University Press, 2016.

Graham Harman, *Object-Oriented Ontology: A New Theory of Everything*, London, UK: Penguin Books, 2018.

Graham Harman, *Art and Objects*, Cambridge UK: Polity Press, 2019.

#62

# Ev

Evolutionary

# #62 An Evolutionary Theory of Design[ing]

*Mark Roxburgh*

Evolutionary theories seek to understand the processes and outcomes of change in the 'natural' world. Design theories seek to understand the processes of human planned change, the outcome of which is the 'artificial' world. Some design theorists see the processes of evolution and design as analogous and the former useful in understanding the latter in terms of how artefacts or families of artefacts come about. Langrish (2004) believes the analogy only useful in understanding the evolution of ideas (memetics) about design. Tonkinwise (2005) argues they are not analogous and conflating them equates to eugenics. It is worth considering, however, that as humans are an 'outcome' of the processes of change in the 'natural' world then our ability to design the 'artificial' world is in essence 'natural' and part of our evolutionary trajectory. In the Anthropocene, design is evolution.

### Notes and Further Reading

John Z. Langrish, "Darwinian Design: The Memetic Evolution of Design Ideas", *Design Issues*, 20: 4, 2004, pp. 4 - 19.

Cameron Tonkinwise, "Design + Evolution = Eugenics: Mimetological Analogies, or Why is Design so Enamoured with Evolution?", *Proceedings of the European Academy of Design EAD06 Conference*, Bremen, Germany, 2005.

#63

# Un

Uncertain

# #63
# An Uncertain Theory of Design[ing]

*Mark Roxburgh*

There might be the possibility that the only thing that might be certain about design is that sometimes our design intentions might not quite turn out as intended and therefore there is a chance that design might be, as Dilnot calls it, 'the science of uncertainty'.

## Notes and Further Reading

Clive Dilnot, "The Science of Uncertainty: The Potential Contribution of Design to Knowledge", in *Doctoral Education in Design: Proceedings of the Ohio Conference*, October 8-11, 1998, (Editor: Richard Buchanan), Pittsburgh: The School of Design, Carnegie Mellon University, 1999.

#64

# Pe

Pessimistic

# #64
# A Pessimistic Theory of Design[ing]

*Mark Roxburgh*

If we take design to be a 'problem solving' activity, as problematic as that concept might be, then our human propensity to design may well end up solving the very problem confronting life on earth as we know it – us and our human propensity to design. In other words, we may well design ourselves, at least in our current form, right out of existence.

## Notes and Further Reading

Nick Bostrom, "Existential Risks: Analyzing Human Extinction Scenarios and Related Hazards", *Journal of Evolution and Technology*, Vol. 9, No. 1, 2002.

R. John Williams, "World Futures", *Critical Inquiry*, Vol. 42, No. 3, 2016, pp. 473 - 546.

#65

# Th

Thoughtless

# #65
# A Thoughtless Theory of Design[ing]

*Mark Roxburgh*

Design thinking is widely regarded as key to innovation and addressing our most complex challenges. However, its thoughtless instrumentalization is implicated in the conception of the surfeit objects cluttering our homes and, when sentiment dictates, storage units, and when it doesn't, the biosphere as landfill, litter and pollution. The image as a representation of a current state we wish to change, an idea of a future state we desire to create, or plan as instruction to manufacture it, is central to design thinking. Design thinking is, therefore, bound up and implicated in the conception of the excess of images cluttering the storage system called the cloud where our sentimentality is perpetually visible, contributing to what Manzini calls "*semiotic pollution*" (Manzini, 1992: 7). It is worth wondering if what we have lost sight of in all this thinking is the imagination to think differently.

## Notes and Further Reading

Ezio Manzini, "Prometheus of the Everyday: The Ecology of the Artificial and the Designer's Responsibility", *Design Issues*, Vol. 9, No. 1, 1992, pp. 5 - 20.

Mark Roxburgh, "The Images of the Artificial or Why Everything Looks the Same", *The International Journal of the Image*, Vol. 3, No. 3, 2013, pp. 1 - 16.

#66

# An

Anthropocentric

# #66
# An Anthropocentric Theory of Design[ing]

*Mark Roxburgh*

As tautological as the concept of human-centred design and its variants such as user-centred design, co-design, and participatory design is, it is at least an honest if unintended admission of the inescapable anthropocentric fact about design. Notwithstanding this fact, our lack of humility as a species in currently privileging this conception overlooks the actuality that we inhabit a complex ecosystem known as the biosphere and that our human-centric approach to design impacts all life forms that inhabit it. Whilst our anthropocentricity is inescapable, we are capable of recognising this and should consider reconceptualising what the centre of design might be to at least mitigate against our anthrochauvinism.

### Notes and Further Reading

Laura Forlano, "Posthumanism and Design", *She Ji: The Journal of Design, Economics, and Innovation*, Vol. 3, No. 1, 2017, pp. 16 - 29.

Paul A. Rodgers and Craig Bremner, "A is for Anthropocene: An A–Z of Design Ecology", *LA+ Interdisciplinary Journal of Landscape Architecture, LA+ Design*, Spring 2019, pp. 110 – 115.

Gerry Scullion, "John Thackara - Designing for all of life, not just human life", 2018, available at: https://www.thisishcd.com/episodes/32-john-thackara-designing-for-all-of-life-not-just-human-life/

Sjoerd van Tuinen, "Air Conditioning Spaceship Earth: Peter Sloterdijk's Ethico-aesthetic Paradigm", *Environment and Planning D: Society and Space*, Vol. 27, 2009, pp. 105 - 118.

#67

# An

Anecdotal

# #67
# An Anecdotal Theory of Design[ing] (or a theory about design theories)

*Mark Roxburgh*

Theories about the natural world (science) are descriptive of what has been repeatedly observed and confirmed through rigorous testing and notwithstanding the instruments and protocols used are human constructs it is generally agreed they produce models of an empirical, verifiable, existing reality. Theories about the artificial world (design) are prescriptive in that they are long-held views established over time, through experience and rhetoric, about the nature of a practice that transforms reality and speculates on what that might be. From Morris' theories about design form and social good to the Bauhaus' theories of design efficiency and function; to more recent theories of design problem-solving and design thinking, theories of design are contested and change, yet their antecedent and antithetical theories are often enmeshed within them. In this regard, design theory is ideological, anecdotal, and tautological.

## Notes and Further Reading

Magdalena Droste, *Bauhaus 1919 - 1933*, Köln, Germany: Benedikt Taschen; 1990 (1st edition).

William Morris, *News from Nowhere*, Oxford, UK: Oxford University Press, 2003 (new edition).

#68

# Ju

Jugaad

# #68
# A Jugaad Theory of Design[ing]

*Saurabh Tewari*

Through a social construction of design beyond the institutional frameworks of modern design, the developing world has catered to its needs with its indigenous ways and expressions. With the constrained economic and material resource scenario, people have been quick-fixing the everyday problems. With its methodical cousins 'gambiarra' in Brazil and 'jua kali' in Kenya, 'jugaad' is one such design approach from India which does more with less. In the rural hinterlands of northern India, a diesel generator turned carrying-cart is one such example which embeds the verb of 'jugaad' as a noun. On the one hand, it exhibits the values of low-resource, makeshift, unstructured innovativeness in people's design. On the other, it can easily be illegal, unsafe, unsustainable, or short-sighted. This duality remains the existential feature of a 'jugaad' theory of design.

### Notes and Further Reading

Anil K. Gupta, *Grassroots Innovation: Minds On The Margin Are Not Marginal Minds*, Gurgaon: Penguin Random House India, 2016.

Navi Radjou, Jaideep Prabhu, Simone Ahuja, Kevin Roberts, *Jugaad Innovation: Think Frugal, Be Flexible, Generate Breakthrough Growth*, New York, NY: John Wiley, 2012.

#69

# Ga

Gandhian

# #69
# A Gandhian Theory of Design[ing]

*Saurabh Tewari*

Mahatma Gandhi, a twentieth-century icon from India, with the methods of ahimsa (non-violence) and swadeshi (from own land), politically catalysed and won the independence war against the colonisers. However, beneath the celebrated political activism supported by communicative ascetic artefacts like charkha (Balaram,1989), lies his empathetic ideas related to social design. Inspired by Ruskin's 'Unto this Last' calling for the upliftment of the weakest, Gandhi's 'Sarvodaya' is a concept which advocates the advancement of all beyond the frameworks of traditional material economics and unjust social structures. With utopian-appearing concepts such as village self-rule, bread labour, critical industrialisation, and decentralisation, Gandhian design ideas are righteous, simple, empathetic, contextual and innovative (Tewari *et al.*, 2017). Khadi, the hand-spun cotton cloth and the process of its making, exhibit design principles which compassionately addresses the poorest of poor, society, nature and humanity at large.

## Notes and Further Reading

Saranya Balaram, "Product Symbolism of Gandhi and Its Connection with Indian Mythology", *Design Issues*, 5 (2), 1989, pp. 68–85, Retrieved from http://www.jstor.org/stable/1511515

Mahatma K. Gandhi, *Ruskin's Unto This Last: A Paraphrase*, Ahmedabad: Navajivan Publishing House, 1956.

Saurabh Tewari, Nipun Prabhakar and Saurabh Popli, "A Gandhian Framework for Social Design: The Work of Laurie Baker and Hunnarshala", In *Smart Innovation, Systems and Technologies* (Vol. 66), 2017, https://doi.org/10.1007/978-981-10-3521-0_29_

#70

# Ra

Rasa

# #70
# A Rasa Theory of Design[ing]

*Saurabh Tewari*

Rasa is a Sanskrit word indicating the essence and emotional abstraction in the classical forms of communication design: visual, literary, musical and performing arts. Ancient India's sage Bharata Muni theorised the aesthetic concept of rasa in the performing arts manual, Natyashastra. He illustrates essential human emotions as shringara (love), haasya (laughter), raudra (fury), karuna (compassion), vibhatsa (disgust), bhayanak (horror), veer (heroism), adbhuta (wonder) and shant (peace). Employed illustratively in the Indian classical performing arts, such as Bharatanatyam and Kathakali, rasa as an aesthetic phenomenon transcends from the creator (performer) to the receptive and sensitive audience. Rasas can augment the experience by enhancing parameters like taste, mood and ambience through semantical elements. In the Indian cultural continuum, rasa theory of design entails not just in the classical arts, but in a broader range, from spirituality to the everyday design encounters.

## Notes and Further Reading

Mathew A. Foust, "The Bloomsbury Research Handbook of Chinese Philosophy and Gender (Book Review)", *Journal of Chinese Religions*, Vol. 45, Issue 1, pp. 108-111, 2017, https://doi.org/10.1080/0737769x.2017.1299410

Susan L. Schwartz, *Rasa: Performing the Divine in India*, New York, NY: Columbia University Press, 2004.

#71

# PI

**Pluralistically Confusing**

# #71
# A Pluralistically Confusing Theory of Design[ing]

*Paul Rodgers & Craig Bremner*

Singular clarity is what we get from despots and politicians - "*...my deal or no deal*" (Parker, 2018). And we know from experience where singular clarity gets us. Nowhere. In terms of design give me pluralistic confusion over singular clarity any day. But three decades ago, Hal Foster (1985) painted pluralism as anything-goes relativism that tends to absorb argument and is uncritical, and this has stuck. More recently, Boris Groys (2016) explained that in pluralistic society, everyone is different but not from each other; different from oneself by refusing to be identified according to the general criteria of identification. Design has built its identity on making one product or process different from another without ever realising that we confuse the identity of design with the need to appear different. Design's plurality, be that the manner in which different disciplines mix with one another, the wildly differing agendas, aims and motivations that designers work to, and the different design styles, techniques and approaches of designing across different geographical, methodological and conceptual boundaries, is in itself not in its products. Rather than confuse, this plurality enriches design. A pluralistically confusing theory of design is all there is...

## Notes and Further Reading

Hal Foster, RECODINGS: Art, Spectacle, Cultural Politics, Seattle, WA: Washington: Bay Press, 1985.

Boris Groys, "The Truth of Art", *e-flux Journal*, No. 71, March 2016, Available at: https://www.e-flux.com/journal/71/60513/the-truth-of-art/

George Parker, "My deal or no deal, Theresa May tells Conservative critics", The Financial Times, September 17, 2018, https://www.ft.com/content/e7b7e402-ba4d-11e8-8274-55b72926558f

# #72

# An

# Anti-Design Thinking

# #72
# An Anti-Design Thinking Theory of Design[ing]

*Jonathan Ventura & Dina Shahar*

People thought before Tim Brown's and Tom Kelley's books (see below). Yet, you must hand it to IDEO for this magnificent contribution to the world of design. However, 10 years later, save for the first-class plane tickets and countless impressive diagrams across the Web, what is left of this concept except for a brilliant and catchy name? What does it really mean, and is it still relevant? During the last decade, we realized that psychology and anthropology should get inside the design studio and influence the process. We even understood that objects may or may not have a unique ontology (yes, it's real, Google it, and no, nobody understands what it truly means). However, following the social, political and economic turmoil of the last decade, should we really focus on "transforming organizations"? A true and significant innovative turn stems from design theory, encompassing, research-based and people-oriented. No buzz-words or fancy diagrams, just good old multi-disciplined... well – thinking.

## Notes and Further Reading

Tim Brown, Change by Design: How Design Thinking Creates New Alternatives for Business and Society, New York, NY: Harper Collins, 2009.

Tom Kelley, The Art Of Innovation: Lessons in Creativity from IDEO, America's Leading Design Firm, London, UK: Profile Books, 2016.

#73

# Bi

Bipolar

# #73
# A Bipolar Theory of Design[ing]

*Jonathan Ventura & Dina Shahar*

Design, as a discipline, suffers simultaneously from inherent inferiority regarding its image, combined with exaggerated prestige, while its practitioners struggle to define their place in the world. At the end of the 20th-century design was regarded as a nice addition, supporting what really matters – engineering, marketing etc. The end of that century saw the rise of a minority of designers who gained super-stardom, which infused their creations with consumption magic. The modern evolution of the discipline bred concepts and jargon which were supposed to infuse the discipline with an air of intellect and importance. Central among these is design thinking. This concept is a flamboyant mixture of inferiority and grandeur. It is an apologetic concept heralding the discipline's importance, while primarily serving marketing purposes. An omnipotent tool for the use of CEOs and organizations.

## Notes and Further Reading

Anthony Dunne and Fiona Raby, *Speculative Everything: Design, Fiction, and Social Dreaming*, Cambridge MA: The MIT Press, 2013.

Johan Redström, *Making Design Theory*, Cambridge MA: The MIT Press, 2017.

#74

# Pr

Proud

# #74
# A Proud Theory of Design[ing]

*Jonathan Ventura & Dina Shahar*

Design theoreticians should not be exiled to a different building! In the last few years, practitioners have come to realize, that contrary to marketing-oriented design, design that makes a meaningful change is embedded in rich, broad, and professionally articulated theories. Future design theoreticians should be immersed in a subject, rather than the old-fashioned history-theory segregation. This original approach will help trigger innovation through multi-discipline professionals based on a specific subject. Healthcare design, for example, will be based on health studies, coupled with medical anthropology and will be practice-oriented targeted at a design situation, be it 2D or 3D, rather than a product. Harnessing a broad array of theories, ranging from ancient philosophy, through haiku poems to social psychology, will create a new form of innovation. Ideation, therefore, will be identified by a mood board, shoulder-to-shoulder with a wall of paragraphs from various thinkers.

## Notes and Further Reading

Patrick Devlieger, Frank R Rusch and David Pfeiffer, *Rethinking Disability: The Emergence of New Definitions, Concepts and Communities*, Antwerp, Belgium: Garant, 2003.

Graham Pullin, *Design Meets Disability*, Cambridge, MA: The MIT Press, 2009.

#75

# St

Studio-Based

# #75
# A Studio-Based Theory of Design[ing]

*Jonathan Ventura & Dina Shahar*

The dreary days of a single, long and tedious theoretical day in your run-of-the-mill design classroom must come to a stop. Employing an unemployed philosopher, art historian or sociologist to teach the same version of "Something 101" s/he remembers from when s/he still had some hope must end. Instead of viewing theory as a way to fill in blank spaces in the design school curriculum, we should reconsider the multi/inter/cross-disciplinary approaches to re-think the still dominant segregation of studio-based learning and theory studies. First, design theory and history should be directly linked to studio projects. While design students are constantly being referred to the design of existing projects, they are hardly being asked to address the cultural, moral, social and economic implications of historic case studies and theoretical approaches relating to their projects. With the segregation of theory and practice, critical thinking is left outside the studio.

## Notes and Further Reading

Jonathan Ventura and Jo-Anne Bichard, "Design Anthropology or Anthropological Design? Towards 'Social Design'", *International Journal of Design Creativity and Innovation*, 5 (3-4), 2017, pp. 222 - 234.

Jonathan Ventura and Galit Shvo, "I Just Cut My Finger in a Ninja Fight: The Semiotics and Hermeneutics of the Band-Aid", *Punctum*, 4 (1), 2018, pp. 179 - 201.

#76

# In

Inclusive

# #76

# An Inclusive Theory of Design[ing]

*Jonathan Ventura & Dina Shahar*

Imagine the future in 10 years. Autonomous, novel transport vehicles; all systems, services and interfaces are personally tailored through big data and deep learning, affordable and friendly robots at our service. This (already cliché) vision is shattered in the face of all the poorly designed interfaces and automated services, leaving us with the question: can we trust designers to consider us, as this fantastic, seamless world materializes in front of our eyes? What can we expect in 2030, if the primitive ATM of 2020 remains obscure and inaccessible, at least for some of us? When first presented, 'Human-Centred Design' sounded important and innovative. But what was the alternative? Aren't all designers participating in designing for humans? As Dieter Rams so eloquently put it - 'the one and only cardinal sin in design is indifference towards people and how they live' (Lovell, 2011). Do contemporary designers follow the legacy of their admired guru?

## Notes and Further Reading

Matteo Bianchin and Ann Heylighen, (2017) "Fair by Design. Addressing the Paradox of Inclusive Design Approaches", *The Design Journal*, 20, Supp 1, pp. S3162 - S3170.

Sophie Lovell, *Dieter Rams: As Little Design As Possible*, London, UK: Phaidon Press, 2011.

#77

# In

## Inflation of Design

# #77 An Inflation of Design Theory of Design[ing]

*Jonathan Ventura & Dina Shahar*

Does building an IKEA closet (if you succeed) make you a designer? On the same note, does visiting a design event (let us say in the northern part of Italy) make you a design theoretician? The answer must be a resounding no! And yet, setting aside the format of this volume, the contemporary reality in which we live, dictates very short and simple texts, almost apologetic for making you use your brain too much. While design theory must not adhere to fashionable trends of oversimplification and force-feeding readers, we must follow several changes. Design theory must be easily understood and readable, even God-forbid – enjoyable (we are talking to you, French thinkers!). As designers will understand that theory is the basis for innovative and meaningful practice, so must theoreticians change and work with and for designers.

## Notes and Further Reading

Henry Dreyfuss, *Designing for People*, New York, NY: Skyhorse Publishing Inc., 2003.

Tomás Maldonado, *Design, Nature, and Revolution*, Minneapolis, MN: Minnesota University Press, 2019.

#78

# Id

Idealist

# #78
# An Idealist Theory of Design[ing]

*Jonathan Ventura & Dina Shahar*

While some of the smugness with which architects address designers stems from almost 2000 years of architectural theory, both disciplines fall under the same shortcoming – the fear of ideology. 'I have nothing to do with the workers' claimed the late Zaha Hadid regarding the horrific deaths of immigrant workers in Qatar, where her al-Wakrah stadium is under construction (Wainwright, 2020). The stadium is 'signature architecture', just like signature chairs by star designers. For many people, this is what design is all about, as ideologically-rooted design projects tend to stay outside the limelight. Why is it so easy, in the few instances we discuss ideology to turn to ecology? Why is the only volume that comes to mind is the same one for the last 50 years? Yes, Victor Papanek's "Design for the Real World". Without ideology, we can't expect care and empathy from designers, and we can all be branded as ideological capitalist pigs.

## Notes and Further Reading

Victor Papanek, *Design for the Real World: Human Ecology and Social Change*, New York, NY: Pantheon Books, 1972.

Oliver Wainwright, "The Despot Dilemma: Should Architects work for Repressive Regimes?", *The Guardian*, Monday 27 January 2020, https://www.theguardian.com/artanddesign/2020/jan/27/despot-dilemma-should-architects-work-for-repressive-regimes-bjarke-ingels

https://www.archdaily.com/480990/zaha-hadid-on-worker-deaths-in-qatar-it-s-not-my-duty-as-an-architect

#79

# Ir

Irrelevant

# #79
# An Irrelevant Theory of Design[ing]

*Joyce Yee*

There are many different ways to live, why choose this one? If we flip this to: there are many ways to use design, why choose this one? Arturo Escobar in his book, "Designs for the Pluriverse" (2018) calls for an urgent refiguring of current design practices, away from feeding the global capitalist machine in a false pursuit of 'modernity' in favour of more collaborative and place-based approaches. There is a growing chorus of critical voices showing discontent with our inability as a community to shift design's infatuation with and slavery to market-driven needs. When are we going to start abandoning design theories that do not offer us a pluralistic, situated and relational view of design? Designers and design educators need to start believing in different ways of living and in practising design – why have theories at all if they are flawed and unfit for the changes ahead?

## Notes and Further Reading

Arturo Escobar, *Designs for the Pluriverse: Radical Interdependence, Autonomy, and the Making of Worlds*, Durham, NC: Duke University Press Books, 2018.

Cameron Tonkinwise, "Design Studies – What Is it Good For?, *Design and Culture*, Vol. 6, Issue 1, pp. 5 – 44.

# #80

# Ev

## Everything and Nothing

# #80
# An Everything and Nothing Theory of Design[ing]

*Mashal Khan*

The increasing categorization of design - speculative, minimalistic, grunge, informative, participatory, sustainable, violent, peaceful, just, reflexive - is problematic. Design is large; it contains multitudes. It is a reflection of the human heart. It is complex, messy, contradictory and in the incompleteness lies the beauty of things modest, humble and unconventional. Design is this constant state of flux, of a work in progress that shape shifts with contextual needs. Words of John Paul Lederach, International Peacebuilding Professor, bespeak design's multidimensionality "*...to envision and respond to the ebb and flow of social conflict as life-giving opportunities.*" (Lederach, 2003: 14). Holding the space of tension, of transformation is when design is everything. Design is nothing when it's too prescriptive and does not address all levels of the problem. Rigidness of design limits possibilities. Design is everything and nothing; things are constantly in a state of becoming or devolving to nothingness. We have a choice: (i) rumble with complexity or (ii) reckon within the specialist's paradigm.

### Notes and Further Reading

John Paul Lederach, *The Little Book of Conflict Transformation*, California, USA: Good Books, 2003.

John Paul Lederach, *The Moral Imagination: The Art and Soul of Building Peace*, Oxford, UK: Oxford University Press, 2005.

#81

# Im

Immersive

# #81
# An Immersive Theory of Design[ing]

*James Fathers*

As someone who thinks that context is everything, I can't imagine a way of designing that doesn't include a complete immersion in the context in which one might be proposing design solutions or suggestions. I suppose one way to explain this would be to describe the opposite approach; to design in a vacuum with no sense of who might use your designs, where they might be used or what impact their use might have on others or the environment. Designers, in my view, must first commit themselves to understand their potential impact in the world, and then keep a check on their egos, so that they only suggest design solutions when they know from their immersive understanding of the given context that there is a real need for a design, an improvement, some 'good change'. It is not clear who said '...the road to hell is paved with good intentions' but allow me to paraphrase. The path to so-called 'progress' is indeed paved with well-meaning ill-informed good intentions. It is time we designers spent more time asking 'why?', 'where?' and 'how?' and 'for who?' before we act.

### Notes and Further Reading

Robert Chambers, *Whose Reality Counts? Putting the First Last*, London, UK: Practical Action Publishing, 1997.

Renwick Irvine, Robert Chambers and Rosalind Eyben, *Learning from Poor People's Experience: Immersions, Lessons for Change in Policy & Organisations, No. 13*, Brighton, UK: Institute of Development Studies, 2004.

John Monaghan and Peter Just, *Social and Cultural Anthropology: A Very Short Introduction*, Oxford, UK: Oxford University Press, 2000.

M.N. Srinivas, A.M. Shah and E.A. Ramaswamy (Editors), *The Fieldworker and the Field: Problems and Challenges in Sociological Investigations*, Delhi, India: Oxford University Press, 1979.

#82

# Fa

Failing Forward

# #82 A Failing Forward Theory of Design[ing]

*James Fathers*

The iterative approach to design is long established in the canon of design law, but allow me to unpack this a little. An iterative approach to design in my view, by definition must embrace a positive attitude to failure. Tom Peters, a business writer, coined the phrase 'failing forwards' in his book "Thriving on Chaos" (Peters, 1989). One of the main challenges design educators face is that students are educated in school to avoid failure at all costs, chasing an 'A' grade almost without thought for the route they have taken. The implication for the design industry is clear because, of course most designers are a product of design education. Unless designers learn to gleefully embrace the joys and rich learning that can come from abject and glorious failures they will risk wallowing in a world of mediocrity where they have avoided risk and taken the safe road to addressing the brief. Ironically, we use the 'lightbulb moment' as a metaphor for inspiration. It apparently took Edison 1000 failures to perfect a working light bulb.

### Notes and Further Reading

Jay Dixit, "Paula Scher on Failure", *Psychology Today*, May 21, 2009, Available at: https://www.psychologytoday.com/us/blog/brainstorm/200905/paula-scher-failure

Tom Peters, *Thriving on Chaos: Handbook for a Management Revolution*, London, UK: HarperBusiness, 1989.

#83

# Tu

Tuning

# #83
# A Tuning Theory of Design[ing]

*Heather Wiltse*

The design of digital, networked, computational things does not result in solid, stable objects, and it does not use up material in the usual sense. It is a process of assembling; of bringing together different components and setting them up to work together. The forms that these things take vary dynamically. The functions they perform are multiple, as are their users - one's use is another's data about use and users. All design is composition, but design of fluid assemblages (Redström and Wiltse, 2019) is permanently compositional without final resolution into totalities (despite appearances). Fluid assemblages do not sit still. They not only come into relations and assemblages, as all things do, but are ongoing relations and assemblages. Designing them entails tuning formations - calibrating functional relations between elements and their collaborative evolution over time. Fluid assemblages are now often characterized by a dissonance between value for end-users and value for producers and others in larger assemblages. A major challenge now is to develop practices of design as tuning formations of fluid assemblages that can harmonize relations of mutual integrity, meaningful transparency, and democratic innovation.

### Notes and Further Reading

Johan Redström and Heather Wiltse, *Changing Things: The Future of Objects in a Digital World*, London, UK: Bloomsbury Publishing, 2019.

#84

# Lu

Ludic

# #84
# A Ludic Theory of Design[ing]

*Emmanuel Tsekleves*

Play (aka Ludus) is one of the main drivers of human cultural development. Play has been the topic of research across different areas including psychology, learning, child development, marketing, Human Computer Interaction and game design. Play and playfulness provides a 'magic circle' (Huizinga, 1955) that fosters the required conditions to encourage exploration and creativity, stimulates social interaction and competition. Play has the potential to foster greater motivation by satisfying three fundamental human needs - the need for competence, autonomy and relatedness. Growing evidence from research reveals that playfulness could serve to promote physical and emotional health, and subjective wellbeing as well as greater levels of physical activity in later life. Playful design, the mapping of playful experiences from games/toys to other non-game context experiences, has gained interest in recent years.

## Notes and Further Reading

Johan Huizinga, Homo Ludens: A Study of the Play-Element in Culture, Boston, MA: The Beacon Press, 1955.

Victor Turner, *From Ritual to Theatre: The Human Seriousness of Play*, New York, NY: Paj Publications, 1982.

#85

# Sp

Speculative

# #85
# A Speculative Theory of Design[ing]

*Emmanuel Tsekleves*

Speculative design is an approach that enables us to think about the future prospectively and critically (Sterling, 2005; Dunne and Raby, 2014). The prime objective of speculative design is to force an aspect of the future into the present so that it demands a response. One of its principal assumptions is the negation of the status quo and the initiation of discussion of possible future scenarios through a confrontation with a tangible object or process, which provides an ideal vehicle for establishing a technology's necessity, its viability and its benevolence within society (Kirby, 2010). It creates narratives for these possible futures in a variety of formats – provocations, prototypes, products, images, films and so on – to express the urgency of change which is required and to focus debate around the actions that could be taken. In these ways, speculative design can be employed to explore citizens' views, concerns and generate debate on policy issues advocated by governments and to genuinely reimagine the relationship between government, community groups and the public.

## Notes and Further Reading

Anthony Dunne and Fiona Raby, *Speculative Everything: Design, Fiction, and Social Dreaming*, Cambridge, MA: The MIT Press, 2014.

David Kirby, "The Future is Now: Diegetic Prototypes and the Role of Popular Films in Generating Real-world Technological Development", *Social Studies of Science*, Vol. 40, Issue 1, 2010, pp. 41-70.

Bruce Sterling, *Shaping Things*, Cambridge MA: The MIT Press, 2005.

#86

# GI

Global Health

# #86 A Global Health Theory of Design[ing]

*Emmanuel Tsekleves*

Ever since the King's Fund Hospital Bed project began in 1962, designers have been undertaking projects for health, and architects have been designing hospitals ever since these institutions were established. Traditionally designers have paid particular attention to acute and chronic care, through new medical products, prostheses, hospital, clinic and care home design. More recently the complex picture of maintaining population wellbeing, of health (ill-being) prevention has begun to emerge, and thus the role of designers indirectly in supporting the promotion of healthy lifestyle or in their contribution to ill-being. As both the cost of healthcare delivery and the ageing population are increasing design is called upon to address such challenges. The issue is how to reduce the cost and burden of disease, particularly of non-communicable disease, by focusing more research work around prevention and looking at how design can work in prevention.

## Notes and Further Reading

Rebecca Partridge, "Understanding the roles of the designer in health care: A Practice-Based study into supporting adolescents with long-term conditions", *The Design Journal*, 20 (4), 2017, pp. 523 - 532.

Emmanuel Tsekleves and Rachel Cooper (Editors), *Design for Health (Design for Social Responsibility Series)*, London, UK: Taylor & Francis, 2017.

#87

# Co

Constructivist

# #87 A Constructivist Theory of Design[ing]

*Mark Roxburgh*

Design might be considered the constructivist discipline par excellence. Our understandings of the processes of designing and the meanings we make of the artefacts we produce subsequent to those processes and the relationships we have with them – our ideas about the 'realities' of design - are socially situated and constructed. Furthermore, the artefacts that get produced (constructed), are often materially situated and experienced and transform the actuality of our material reality. If largely implied in the design literature, work in the field of the sociology of technology, such as actor network theory, provides extensive empirical and theoretical evidence to demonstrate these points. As the epistemology and ontology of design is constructivist, it is also relativist because its social and material contingency is mutable. With relativism comes the risk of fundamentalism. Perhaps this might be the proper study of design fundamentals.

## Notes and Further Reading

Wiebe E. Bijker and John Law (Editors), *Shaping Technology / Building Society: Studies in Sociotechnical Change*, Cambridge, MA: The MIT Press, 1992.

Ranulph Glanville, "Researching Design and Designing Research", *Design Issues*, Vol. 15, No. 2, 1999, pp. 80 - 91.

Donald MacKenzie and Judy Wajcman (Editors), *The Social Shaping of Technology*, Milton Keynes: Open University Press, 1985.

#88

# Ca

Caring

# #88
# A Caring Theory of Design[ing]

*Jen Archer-Martin*

Caring design theories and practices intra-act to make a difference in the world, bringing care to practices of doing and thinking design. Whether informed by phenomenological care-as-being (Heidegger, 1978), feminist ethics of care (Gilligan, 1982; Tronto, 2012), indigenous knowledges of caring interrelations, emerging understandings of care for the more-than-human, or simply an intuitive understanding of what it means to care and be cared for, caring design theory nurtures the capacity of design, and its communities, to care. It recognises design and theory as entangled practices operating from within a given, situated ecology, drawing attention to the potential of caring relations within that ecology to strengthen the vitality of the system and its constituents. Antithetical to an abstract set of universal laws, a caring theory of design is an embodied, living practice of caring for, about, with and through design.

## Notes and Further Reading

Jen Archer-Martin, "Magical Agents: The Powers of Care are not Ours Alone," *Does Design Care…?*, Lancaster, UK: Lancaster University, 2017, pp. 113-120.

Karen Barad, *Meeting the Universe Halfway: Quantum Physics and the Entanglement of Matter and Meaning*, Durham, NC: Duke University Press, 2007.

Berenice Fischer and Joan Tronto, *Towards a Feminist Theory of Care. Circles of Care: Work and Identity in Women's Lives*, Albany, NY: State University of New York Press, 1991.

Carol Gilligan, *In a Different Voice: Psychological Theory and Women's Development*, Cambridge, MA: Harvard University Press, 1982.

Carol Gilligan, "Hearing the Difference: Theorizing Connection", Hypatia, 10, No. 2, 1995, pp. 120-127.

Martin Heidegger, *Being and Time*. Oxford: Basil Blackwell, 1978.

Maria Puig de La Bellacasa, *Matters of Care: Speculative Ethics in More Than Human Worlds*, Minneapolis, MN: University of Minnesota Press, 2017.

Joan C. Tronto, "Partiality Based on Relational Responsibilities: Another Approach to Global Ethics", *Ethics and Social Welfare*, 6 (3), September 2012, pp. 303–316.

#89

# Mo

## More-than-Human(ist)

# #89
# A More-than-Human(ist) Theory of Design[ing]

*Jen Archer-Martin*

Recent decades have seen much touting of humanist values in design, from human-centred design's focus on the experience of using design products, to co-design's engagement of users in the processes of design. While design must still work to embrace the plurality, diversity and fluidity of human agency, it is also confronted with the more-than-human. Augmenting existing ethical practices is an evolving awareness of the non-human others with which we co-inhabit this planet (and beyond), and the interconnectedness of our survival. A new agenda for design is being set by more-than-human (including post-anthropocentric, post-human, new materialist and ecological) discourse. Where humanist design is socially engaged, more-than-humanist design is ecologically engaged. Further to considering design's impact on ecologies, this agenda asks: Who/what are the animate/inanimate, natural/technological agents of more-than-human design? How might they be engaged in design processes and experiences?

## Notes and Further Reading

Richard Buchanan, "Human Dignity and Human Rights: Thoughts on the Principles of Human-Centered Design", *Design Issues*, Vol. 17, Issue 3, 2001, pp. 35 - 39.

Joseph Giacomin, "What Is Human Centered Design?", *The Design Journal*, Vol. 17, Issue 4, 2014, pp. 606 - 623.

David Kelley, "Human-Centred Design," filmed February 2002 in Monterey, CA. TED video, 16:57, https://www.ted.com/ talks/david_kelley_on_human_centered_design.

Ezio Manzini, "Design Culture and Dialogic Design", Design Issues, Vol. 32, Issue 1, Winter 2016, pp. 52 - 59.

#90

# No

Non-Cartesian

# #90
# A Non-Cartesian Theory of Design[ing]

*Peter Lloyd Jones*

Exponential leaps in computation power have resulted in the emergence of a new kind of practice in which design is no longer based upon history, or a priori knowledge of form, materiality, technology, or engineering. For this theory of design, experimentation and form-finding are the first step in a 'Search and Find' mass data retrieval and filtering exercise that precedes rapid rounds of simulation and optimization in silico, replacing logic, deduction, experience, and centuries-old Cartesian principles rooted in the mathematical and physical sciences. For this "Second Digital Turn", Mario Carpo argues that: "*...the apparently unfathomable complexity of the physical shapes they are now creating already expresses a new form of artificial intelligence, outside the tradition of modern science and alien to the organic logic of our mind.*" (Carpo, 2017: 81). Scale and standards are simply vanishing, as this theory plays out.

## Notes and Further Reading

Armen Avenassien, "Another History of Science, an Interview with John Tresch", *Mousse Magazine*, Issue 37, February - March 2013, pp. 244 - 246.

Mario Carpo, *The Second Digital Turn: Design Beyond Intelligence*, Cambridge, MA: The MIT Press, 2017.

#91

# Ea

Early Learning

# #91 An Early Learning Theory of Design[ing]

*Peter Lloyd Jones*

Collaborations between designers and scientists are increasing, yet cannot reach their potential if efforts aren't made to understand one another. Author and scientist C.P. Snow acknowledged this problem in his legendary 1959 Rede lecture titled Two Cultures. "*Once or twice I have been provoked and have asked the company (of non-scientists) how many of them could describe the Second Law of thermodynamics. The response was cold - it was also negative. Yet I was asking something which is the scientific equivalent of - 'Have you read a work of Shakespeare's?' I now believe that if I had asked an even simpler question, such as - 'What do you mean by mass, or acceleration?', which is the scientific equivalent of saying: 'Can you read?*" (Snow, 1998: 14-15). An early learning theory finds commonalities between design and science, beginning with A is for Art as a shared tool.

## Notes and Further Reading

Paul A. Rodgers and Craig Bremner, *Design School: After Boundaries and Disciplines*, Malaga, Spain: Vernon Press, 2019.

C.P. Snow, *The Two Cultures,* Cambridge, UK: Cambridge University Press, 1998.

#92

# Sc

Scientific

# #92
# A Scientific Theory of Design[ing]

*Peter Lloyd Jones*

Science tells us that prospecting represents a unique human survival factor, which allows us to imagine and navigate our way to a better place or state of being in the future. Design must be a way by which we get there. In other words, to design is to be human; a fact so fundamental that it predates any conceit of a stand-alone discipline. Logic also dictates then, that if design exists in the future, neither the current spaces used to focus it nor the names used to frame it, currently exist either. By extension, if we accept that design is a nameless, non-discipline that exists in the future within an imagined, but currently unknowable space, where it is made real, then it's plausible to speculate upon a theory of design that is boundless and pre-disciplinary, just like science.

## Notes and Further Reading

Peter Lloyd Jones, "Minding the Gap: Caring for Collaboration between Art & Science by Design", *SciArt Magazine* (digital edition/SciArt Initiative Inc), 2018, available at: https://www.sciartmagazine.com/minding-the-gap-caring-for-collaboration-between-art--science-by-design.html#

Martin E. P. Seligman, Peter Railton, Roy F. Baumeister and Chandra Sripada, *Homo Prospectus*, New York, NY: Oxford University Press, 2016.

#93

# Ca

Catastrophic

# #93
# A Catastrophic Theory of Design[ing]

*Peter Lloyd Jones*

Theories shared between design and science can save millions of lives. Inspired by the works of Buckminster Fuller and Kenneth Snelson, for example, "Tensegrity" has been successfully imported from design into biology to help explain how form controls function in living systems, resulting in the design of novel and more effective medical treatments for diseases including cancer. Similarly, production of revolutionary gene editing tools within the new design field of synthetic biology, which includes CRISPR/DNA editing methods, are rapidly changing the clinical management of multiple human pathologies, while simultaneously changing the rules as to how non-biological things are designed, made, and organized. Recent mathematical models, however, indicate that the off-target effects of CRISPR may produce unintended, negative effects on human health, including damage to genes and ecosystems, indicating that synthetic life may actually represent a catastrophic theory of design.

## Notes and Further Reading

Jon Cohen, "Did CRISPR help—or harm—the first-ever gene-edited babies?", *Science Magazine*, August 1, 2019, available at: https://www.sciencemag.org/news/2019/08/did-crispr-help-or-harm-first-ever-gene-edited-babies#

Peter Lloyd Jones, "Context Messaging: Modeling Biological Form", In: 306090 (Models), Vol 11, 2006, pp. 33 - 38 (Princeton University Press).

Jenny E. Sabin and Peter Lloyd Jones, *LabStudio: Design Research Between Architecture and Biology*, New York, NY: Routledge, 2017.

#94

# Di

Disempowering

# #94
# A Disempowering Theory of Design[ing]

*Heather Wiltse*

Given the importance of data for business intelligence, connected products that can produce data (*i.e.*, all of them) should be designed to maximize the production of this resource. This involves ensuring that users do not take steps that can effectively reduce the production of data about their use. Users should not feel that they have any real choice if they would like to continue using services that enable them to connect to other people, create documents and other media, find information, read the news, and generally participate in the practices of contemporary life. They should grant consent to the collection and broadly-defined use of personal data without restrictions in application, scope, or scale. So-called "dark patterns" of interface design should gently guide users toward the light of intended behaviour. In short, design should disempower users when it comes to controlling the ways in which their presence and activities are registered and made visible in the world in the form of data when they use connected things. Or should it?

**Notes and Further Reading**

Nick Couldry and Ulises A. Mejias, *The Costs of Connection: How Data is Colonizing Human Life and Appropriating It for Capitalism*, Stanford, CA: Stanford University Press, 2019.

Forbrukerrådet, "Deceived by Design: How tech companies use dark patterns to discourage us from exercising our rights to privacy", 2018, available at: https://www.forbrukerradet.no/undersokelse/no-undersokelsekategori/deceived-by-design

Shoshana Zuboff, *The Age of Surveillance Capitalism: The Fight for a Human Future at the New Frontier of Power*, London, UK: Profile Books, 2019.

#95

# Fi

Fictional

# #95
# A Fictional Theory of Design[ing]

*Craig Bremner*

In the era of the production of nothing, design is passing through a phase describing its actions as producing fictions – narrative replacing object – giving rise to fictional design that of course misunderstands the functioning of fiction in literature. To Michel Foucault, it seemed "*...that the possibility exists for fiction to function in truth*" (Foucault, 1980: 193). But having made a complete mess of the real-world, design is contriving another world for its petty thoughts and clumsy actions – a fictional world for its speculative design. This paradox is best described by the writer Gerald Murnane as: "*wanting to understand how the so-called actual and the so-called possible (what one did and what one only dreamt of doing) come finally to be indistinguishable in the sort of text that we call true fiction.*" (Murnane, 2014: 140). A fictional theory of design exposes the actual character of design – it is possible to speculate only by evading the truth.

## Notes and Further Reading

Michel Foucault, *Power/Knowledge: Selected Interviews and Other Writings 1972–1977*, edited by Colin Gordon, New York, NY: Pantheon Books, 1980.

Gerald Murnane, *A Million Windows*, Sydney, Australia: Giramondo Press, 2014.

Paul A. Rodgers and Craig Bremner, "The Design of Nothing: A Working Philosophy", in *Advancements in the Philosophy of Design*, edited by Pieter Vermaas and Stephane Vial, Berlin, Germany: Springer Press, 2018, pp. 549 – 564.

#96

# Ex

Excessive

# #96
# An Excessive Theory of Design[ing]

## *Mark Roxburgh*

It is estimated that the average 19th-century family home contained approximately 200 manufactured items, whereas its early 21st century equivalent contained 300,000. Whilst it beggars belief that contemporary life is so complex that 21st century families face a staggering 299,800 more problems requiring design solutions than their 19th-century equivalent, design has nonetheless solved the problem of this increased complexity through ever more ingenious storage solutions, larger houses, multi-car garages (full of stuff, not cars), and the genius of rented storage facilities. Typically, we assume that filling our lives with possessions points to problems of inadequacy or relentless sentimentality. Regardless we can only solve these problems ourselves, most likely without design. Alternatively, Daniel Miller (2009) found that our possessions enhance our relationships with people, not replace them. We can only assume then that human relationships are at an all-time historical high.

### Notes and Further Reading

Daniel Miller, T*he Comfort of Things*, London, UK: Polity Press, 2009.

https://www.latimes.com/health/la-xpm-2014-mar-21-la-he-keeping-stuff-20140322-story.html

#97

# Ma

Management

# #97
# A Management Theory of Design[ing]

*Rachel Cooper*

Design has often been an adjunct to another discipline, and when design found management, or rather management found design, it seemed to be the ideal way in which we could bring discipline to an undisciplined discipline, and illustrate its use and value to business and industry. Yet this activity has never had a definitive definition or clarity of purpose. In the Oxford English Dictionary design is defined as "*purpose or planning that exists behind an action, fact, or object*" (mass noun) and management as "*The process of dealing with or controlling things or people*" (mass noun). It would appear that you can't really do one without the other. Indeed, in most situations, whether commercial, political or social, those people who make decisions are making design decisions that affect the products we use, the places we inhabit and the services we receive, these decisions then rely on design and management for effective implementation of the decision.

## Notes and Further Reading

Rachel Cooper and Mike Press, *The Design Agenda: A Guide to Successful Design Management*, Chichester, UK: John Wiley & Sons, 1995.

Rachel Cooper, Sabine Junginger and Thomas Lockwood, *The Handbook of Design Management*, London, UK: Berg Publishers, 2011.

David Hands, *Design Management: The Essential Handbook*, London, UK: KoganPage, 2017.

#98

# De

## Design Thinking

# #98 A Design Thinking Theory of Design[ing]

*Rachel Cooper*

Do designers think? Of course, they do. Is design thinking different from designing? Well, in order to design one should first think. In a recent review of the design literature, the evidence suggests that to design is to communicate, through design practice, design visualisation, design prototypes and methods of researching, thinking, doing and making, to get to these visualisations. This process forms a common ground upon which further conversations can be built. Design thinking is inseparable from design doing. In other words, designing.

## Notes and Further Reading

Nigel Cross, *Design Thinking: Understanding How Designers Think and Work*, London, UK: Bloomsbury, 2011.

Ricardo J. Hernández, Rachel Cooper, Bruce Tether, Emma Murphy, "Design, the Language of Innovation: A Review of the Design Studies Literature", *She Ji: The Journal of Design, Economics, and Innovation*, Vol. 4, No. 3, Autumn 2018, pp. 249 - 274.

Bryan Lawson, *How Designers Think: The Design Process Demystified*, London, UK: Routledge, 2005 (4th edition).

Peter G. Rowe, *Design Thinking*, Cambridge, MA: The MIT Press, 1991 (new edition).

#99

# Me

Messy

# #99
# A Messy Theory of Design[ing]

*Stephanie Carleklev*

Theory is often viewed as the antipode of practice. It is the opposite of getting your hands dirty and creating a mess. But for design, this artificial dichotomy is rather problematic. Design depends on the integration of both, as described by Nelson and Stolterman (2012). Design is to imagine that things can be different – and to turn this thinking into action. Like riding a bike, practising design depends on getting out there, facing your surroundings, other traffic, and the limitations of your body. And while it can result in a bloody knee or two, this immersion also yields surprises and discoveries that could never have been made otherwise. Practice is often a rather invisible, but subsequently no less important negotiation between our imagination and the reality of materials, abilities and relationships. It is the way human life is lived – and design is practised.

### Notes and Further Reading

Tim Ingold, *Making: Anthropology, Archaeology, Art and Architecture*, Abingdon, UK: Routledge, 2013.

Harold G. Nelson and Erik Stolterman, *The Design Way: Intentional Change in an Unpredictable World*, Cambridge, MA: The MIT Press, 2012.

#100

# Ca

Careful

# #100
# A Careful Theory of Design[ing]

*Stephanie Carleklev*

Can we really say that much has changed since Victor Papanek (1985) criticised design as being extremely dangerous? Almost fifty years since the publication of the first edition of 'Design for the Real World' in 1971, design is still largely entangled in a wasteful creation of artefacts and messages, many of us neither need nor care about. And despite numerous attempts to take on responsibility and to demonstrate design's ability to enable change, design has only just begun to define a careful theory of design[ing]. Regardless, the fact that we adore design's ability, it is a process that will gain from balancing pride and ambition with the acceptance of design's limitations. Not just once has design's much-praised ability to imagine preferred futures had the opposite result. We should consider it a warning. Maybe it is time to let go of our dream to predict, to know everything and to bring this world to perfection – ignoring the complexity, multiplicity and uncertainty of reality. A careful theory of design[ing] has fewer answers and embraces the life we care for in a rather inconsistent, inaccurate and unpredictable manner.

## Notes and Further Reading

Clive Dilnot, "Nine Swallows – Perhaps Summer? The Last Two Decades of Design Studies", *The Design Journal*, 20 : 2, 2017, pp. 165 - 180.

Tony Fry, *Design Futuring – Sustainability, Ethics and New Practice*, Oxford, UK: Berg Publishers, 2009.

Victor Papanek, *Design for the Real World: Human Ecology and Social Change*, London, UK: Thames and Hudson, 1985 (2nd edition).

#101

# Ag

Agile

# #101
# An Agile Theory of Design[ing]

*Stephanie Carleklev*

Is design about objects or processes? Is it about what can be touched or looked at, or what is made possible by it? While we can't answer where the line is drawn between design's proposition and a life that is lived irrespectively, design can simply not provide the props for a play without curiously investigating its plot. Design is enacting itself in verbs, in activities made possible. Instead of designing a book, we design reading, imagining, provoking. In the case of a kitchen, we create the affordance (Gibson, 1966, 1979) of cooking, laughing, breaking a glass once in a while – and hopefully a lot more. Acknowledging and engaging with the ever-increasing complexity in which design operates, it becomes more and more impossible to consider design as a static entity in a void. To see design as a process or event that unfolds over time allows it to exceed the limitations of its disciplinary boundaries and to explore the potential of new methods, tools and materials that result from it.

### Notes and Further Reading

James J. Gibson, *The Senses Considered as Perceptual Systems*, London, UK: Allen and Unwin, 1966.

James J. Gibson, *The Ecological Approach to Visual Perception*, Boston, MA: Houghton Mifflin Harcourt, 1979.

Ramia Mazé, *Occupying Time: Design, Technology, and the Form of Interaction*, Stockholm, Sweden: Axl Books, 2007.

#102

# Lu

Ludic

# #102
# A Ludic Theory of Design[ing]

*Cathy Treadaway*

Ludic design playfully challenges accepted convention and stimulates both the mind and the senses. 'Ludic' means playfulness that is open, non-goal directed and exploratory – it avoids rules and stimulates pleasure. All design is playful at some stage, since playfulness is a vital component of creativity – it helps explore possibilities and pushes the boundaries of what is possible. Designers who are playful with their design concepts break the rules, bring surprises and generate humour. The playful process of design may tease and disrupt, but it can also bring a smile – it is purposefully emotional and can play with the heartstrings. Ludic design tests the potential of what is considered acceptable and possible in the market place; for example, Philippe Starck's lemon squeezer for Alessi was inspired by a plate of calamari and designed on a paper napkin. Ludic designs take us into new realms of possibility and fire the imagination.

## Notes and Further Reading

Johan Huizinga, *Homo Ludens: A Study of the Play-element in Culture* (Humanitas, Beacon reprints in Humanities), Boston, MA: Beacon Press, 1955.

Donald A. Norman, *Emotional Design: Why we Love (or Hate) Everyday Things*, New York, USA: Basic Books, 2004.

Mitchel Resnick, "All I Really Need to Know (About Creative Thinking) I Learned (By Studying How Children Learn) in Kindergarten", *Proceedings of the 6th ACM SIGCHI Conference on Creativity & Cognition*, Washington DC, USA, June 2007, pp. 1 - 6.

#103

# Sm

Smart

# #103
# A SMART Theory of Design[ing]

*Delfina Fantini van Ditmar*

'SMART', is usually characterised by having some sort of 'intelligence' but also by being tidy and cold, frequently it presents itself capitalized. The SMART paradigm has landed from the corporate tech world into design. SMART materials // SMART objects // SMART homes // SMART cities// SMART behaviour// SMART performance. As an adjective fastened to every possible domain of our existence, SMART appears to be a contestable imperative for contemporary designers. A SMART theory of design[ing] embraces the uncertainty of what the definition of SMARTness would encompass. A SMART theory of design[ing] is based on critical, relational and contextual thinking. SMARTness constantly questions itself and is very aware that is not neutral. It refuses to systematize humans and reducing us into parameters of optimization, productivity and efficiency. In a SMART theory of design[ing], SMART gets frustrated when it is reduced to digital traits and used purely as a commercial tag. A SMART theory of design[ing] doesn't aspire to techno-solutionism, it questions and challenges algorithmic reductionism. It continuously reconstructs and reframes SMARTness. SMARTness is not a commodity that you can store.

**Notes and Further Reading**

Evgeny Morozov, *To Save Everything, Click Here: Technology, Solutionism, and the Urge to Fix Problems that Don't Exist*, London, UK: Penguin Books, 2014.

Fred Turner, *Counterculture to Cyberculture: Stewart Brand, the Whole Earth Network, and the Rise of Digital Utopianism*, Chicago, USA: University of Chicago Press, 2008.

Judy Wajcman, *Pressed for Time: The Acceleration of Life in Digital Capitalism*, Chicago, USA: University of Chicago Press, 2015.

#104

# 1.5

1.5°C

# #104
# A 1.5°C Theory of Design[ing]

*Delfina Fantini van Ditmar*

Following the Intergovernmental Panel on Climate Change's (IPCC) Fifth Assessment Report, design can't neglect the responsibility that the field has changed on the 1.5 °C above pre-industrial levels. A 1.5°C theory of design[ing] if forested with Co2 interrogations might help. On a warming planet, designers must strive to rethink supporting policies and to act upon consumption culture. Bellow 2 °C requires a high degree of imagination; sustaining levels won't lead to the 1.5°C target. A 1.5°C theory of design[ing], designers must aim for negative emissions. In the geological era of the Anthropocene, the 1.5 °C threshold, is one of the most meaningful challenges that the discipline has ever faced. A 1.5°C theory of design[ing] specifies that the problem is beyond a local political (economic?) consideration; a global commitment of the field is required. It is clear is that designers will need to adjust quickly to the scale of the problem. With no coral reefs, we should prepare for the waves. In a 1.5°C theory of design[ing], designers have no opportunity to evaporate.

**Notes and Further Reading**

The Fifth Assessment Report (AR5) of the Intergovernmental Panel on Climate Change (IPCC), 2014. https://www.ipcc.ch/assessment-report/ar5/

Shannon Mattern, "The Big Data of Ice, Rocks, Soils, and Sediments: Inside the Material Archives of Climate Science", *Places Journal*, November 2017, Available at: https://placesjournal.org/article/the-big-data-of-ice-rocks-soils-and-sediments/

#105

# Cu

Cultural etc...

* While there are no formal references for them, we acknowledge with sincere gratitude Danièle's Elders, Aunty Gloria Nipperess (Budawang/Yuin) and Uncle Greg Simms (Gadigal, Dharug, Gundungurra, Budawang), who patiently over many years and countless instances taught her respectful ways of relating to Country and community using oral methods of transmission.

# #105
# A Cultural Generational Communal Iterative Generative Collaborative Relational Theory of Design[ing]

*Danièle Hromek (Budawang/Yuin) & Jacqueline Gothe*

This theory declines the predominant theories of design[ing] which primarily value the individual, the designer and the object and acquires significance from Western tradition. It acknowledges the deficiency of that tradition to express the truths of First Peoples' worldviews. In this context of colonisation where the subjugation of knowledges, oppression of languages and the contemporary practices of design[ing] create an incongruence that disables the expression of cultural, spiritual, relational and material responsibility in design[ing]. This theory of design[ing] reclaims and claims generative, collaborative and respectful practices that give meaning to the agency of the more-than-human, the land and the relationships that connect everything. This theory of design[ing] emerged through relationships and reconnections reclaiming ancestral, generational, historical, story continuum – past present and future – and connections that hold us in the lived moment of place, time and memory.

### Notes and Further Reading*

Fran Bodkin, *D'harawal Seasons and Climatic Cycles*, Sydney: Envirobooks, 2008.

Mary Graham, "Some Thoughts about the Philosophical Underpinnings of Aboriginal Worldviews", *Australian Humanities Review*, 45 (November), 2008.

Max Dulumunmun Harrison and Peter McConchie, *My People's Dreaming: An Aboriginal Elder Speaks on Life, Land, Spirit and Forgiveness*, Warriewood, NSW: Finch Pubs., 2009.

Ambellin Kwaymullina, "The Creators of the Future: Women, Law and Telling Stories in Country", In P. Dudgeon, J. Herbert, J. Milroy, and D. Oxenham (Eds.), *Us Women, Our Ways, Our World*, Broome, WA: Magabala Books, 2017.

Victor Steffensen, *Fire Country How Indigenous Fire Management Could Help Save Australia*, Melbourne: Hardie Grant, 2020.

#106

# Pl

Plurality

# #106
# A Plurality Theory of Design[ing]

*Nelly Ben Hayoun*

Architecture and design played a key role in the affirmation of ideological and authoritarian regimes where one style ruled it all. In this context, a lot of what we teach at the University of the Underground - the free university I run in the basement of nightclubs - is informed by the philosophy of political theorist Hannah Arendt (1958, 1961), who passed away in 1975 and who predicted the 'alienation of humanity' due to the tendency of the engineer to be sublimated by their technological tools, and often our incapacity to think critically. In the University of the Underground, we are concerned with plurality and future thinking, with classes in linguistics, political philosophy, music, design, theatre, film, technology and science. Students are tasked to develop experiences in the context of the institutions of their choice, public and/or private, to seed change and modify power structures within them. We spend a lot of time thinking about what needs and requires change - more specifically in institutions and how experiential design practices can play a role in that. What Hannah Arendt was advocating for was the plurality of thinking. As opposed to the 'one idea rules it all' of ideology, she thinks in term of plurality. Other philosophers like Guattari or Deleuze, talk about 'the rhizome' and a more schizophrenic approach to linear thinking as a means to engage with a multiplicity of thoughts.

### Notes and Further Reading

Hannah Arendt, *The Human Condition*, Chicago: University of Chicago Press, 1958.

Hannah Arendt, *Between Past and Future*, New York, NY: The Viking Press, 1961.

Gilles Deleuze and Félix Guattari, *A Thousand Plateaus*, Translated by Brian Massumi, London and New York: Continuum, 2004.

#107

# De

Designer-as-
Mythologist

# #107
# A Designer-as-Mythologist Theory of Design[ing]

*Nelly Ben Hayoun*

In 1957, French linguist and philosopher Roland Barthes published a series of 'mythologies' that he extracted from the news between 1954 to 1956. With titles ranging from "The roast beef and the chips" to the "Martians" or the "Red wine", these short notes on "collective representations" were studied as systems of signs. He explored common 'mythologies', and the relationship between language, aspiration and power. Barthes says that myths - which are a big part of the production of our culture - are not communicated officially, they are not visible at first, they have no politics, and they seem to belong to all of us. The difficulty to study, to report or to do a critical study of myths relies on a skilled mythologist. According to Barthes, the mythologist is the one who 'liaises with the world in sarcastic terms' - they/she/he has the ability to maintain a critical distance from mass culture. Indeed, the mythologist has a key role to play - in the way we think of our needs - as truth and facts are now the subjects of intense questioning. The opacity that rules decision-making processes means that whoever can reveal these systems, and render them visible to audiences (the public) has both a clear power to lead organized groups, but also action change within our existing structures. This, I believe, should be the role of the designer.

## Notes and Further Reading

Roland Barthes, *Mythologies*, Paris: Editions du Seuil, 1957.

Roland Barthes, *Mythologies* (translated by Annette Lavers), London: Paladin, 1972.

#108

# On

Ontological

# #108 An Ontological Theory of Design[ing]

*Arturo Escobar*

Design[ing] creates worlds; whether through the microworlds and micro identities spawned by mundane designerly objects (chairs, cups, cellphones), meso-level inventions (cars, homes, buildings), or through large-scale interventions (cities, science, health systems) designers fashion ways of being, thinking and doing. Entrenched quotidian rituals of object use, naturalized suburban practices of habitability, cities detached from Earth, disjointed bodies made up of independent organs enacted by hospitals and MDs... In all of these domains there lurks the design[ing] hand, always ready to ontologically strike. Even romantic love has not been immune to design's allopoietic force. Never forget that in designing, we are designed. Such design frenzy has never been truer than in the modern era, which for this very reason might be, finally, coming to an end. Can we seriously commence the civilizational task of ontologically redesigning our worlds? Who/how shall we build the prototypes of the multiple civilizations of the futures to come?

## Notes and Further Reading

David Bollier and Silke Helfrich, *Free, Fair and Alive: The Insurgent Power of the Commons*, Gabriola Island, BC: New Society Publishers, 2019.

Tony Fry, *Becoming Human by Design*, London, UK: Berg, 2012.

Terry Winograd and Fernando Flores, *Understanding Computers and Cognition*, Norwood, NJ: Ablex Publishing Corporation, 1986.

#109

# Ma

## Materials' History

# #109
# A Materials' History Theory of Design[ing]

*Arturo Escobar*

Like art, design[ing] would learn from taking into account the millenarian history and geography of materials. "*In our era of combine harvesters, we have lost the sense of the wheat ear*", says the great phenomenologist Gaston Bachelard (1970: 23). At their best, artists "*...return us to the grandeur of origins.*" In crafting a piece, the iron sculptor, becoming blacksmith, follows the complex soul of iron, co-evolving with the metal and meditating with the material, caring not about the final result but about the work itself. Tool, hand, and material dream together, as the artists wills it into form. Can design[ing] emulate this poetics of materials? Can it awaken the souls of iron, stone, earth, wood, inks, cellulose, silica? "*Eons ago, in a past that no longer belongs to us, live within us the dreams of the forge. It's healthy to make them come to life again.*" (Bachelard, 1970: 23). Design[ing] must meditate materially on this forgotten history. What do plastics - or so many new materials - know, on the contrary, of the Dream of the Earth?

## Notes and Further Reading

Gaston Bachelard, *The Poetics of Space* (trans. Maria Jolas), Boston, MA: Beacon Press, 1969.

Gaston Bachelard, *Le Droit de Rêver*, Paris, France: Les Presses Universitaires de France, (1re édition), 1970.

David Joselit, Carrie Lambert-Beatty and Hal Foster, "A Questionnaire on Materialisms", *OCTOBER*, 155, Winter 2016, pp. 3–110.

#110

# Co

Communal

# #110
# A Communal Theory of Design[ing]

## *Arturo Escobar*

One of Modernity's most definitive achievements has been the paradoxical production of de-communalized individuals *en masse*. The more we design, often lovingly, our most cherished personal niches - as professionals, citizens, inhabitants of places, even lovers - the more we are designed to design our lives as allegedly independent individuals. Our individualities thus largely become the product of individualizing design forces. But life is interdependence through and through. For each of us to exist, all - everything - must also exist. We are deeply enmeshed in intricate webs of relations. The more quickly, and deeply, we come to the realization that In the Beginning There Was The Relation - it always is - the more thorough will be our resolve to design for reconnection with fellow human and non-human beings. We would finally be recommencing our journey towards communal ways of being. This is a *sine qua non* for healing our fragmented and isolating cultures. A communal theory of design would tackle in earnest this task of ontological repair.

### Notes and Further Reading

Arturo Escobar, *Pluriversal Politics: The Real and the Possible*, Durham, NC: Duke University Press, 2020.

Tim Ingold, *Being Alive: Essays on Movement, Knowledge, and Description*, New York, NY: Routledge, 2011.

Kriti Sharma, *Interdependence: Biology and Beyond*, New York, NY: Fordham University Press, 2015.

#111

# Au

Autonomous

# #111
# An Autonomous Theory of Design[ing]

*Arturo Escobar*

Throughout its celebrated history, modern design has served all kinds of clients. Constituencies of diverse kinds nowadays pile up at the door of fancy design consultancies and firms. Even social innovation becomes a matter of design, and deservedly so. Could we fathom the case of small bands of designers exercising their craft in support of communities in struggle? In much of the Global South, one finds all kinds of communities mobilizing to defend their territories and lives from the ravages of large-scale extraction. Two banners are often found at the forefront of these anti-systemic struggles, as they aim for radically different ways of worlding in the face of the globalizing juggernaut - the reweaving of communal territories, and cultural and political autonomy. Autonomous design - design that takes at its main goal the strengthening of the collective autonomy of the communities - may become a praxis of healing with communities for designers so inclined.

## Notes and Further Reading

Arturo Escobar, *Designs for the Pluriverse: Radical Interdependence, Autonomy, and the Making of Worlds*, Durham, NC: Duke University Press, 2018.

Alfredo Gutiérrez Borrero, "Resurgimientos: sures como diseños y diseños otros", *Revista Nómadas*, 43, 2015, pp. 113 - 129.

Ezio Manzini, *Politics of the Everyday*, London, UK: Bloomsbury, 2019.

#112

# Pl

Pluriversal

# #112
# A Pluriversal Theory of Design[ing]

*Arturo Escobar*

Design has coexisted comfortably with all of the universals of modernity - progress, science, truth, value, markets, and capitalism itself. These universals arrest life, fix it into forms that are presented as natural, as if life really were what the dominant categories assume. Show me a universal, and I will show you a form - many forms - of design breathing at ease in the space it defines. But, if life were a pluriverse, instead of a universe? A world where many worlds fit, as the Zapatista of Chiapas brilliantly state, rather than the 'global village' of consumers imagined by globalization gurus and complacent branding agencies? Is life always ceaseless flow, endlessly reverberating into milliard forms, and so impossible to cage into structures and routines? If so, design could become the task of mindfully and effectively weaving the mesh of life, the pluriverse of worlds and forms that modernity has sought to banish with its universalizing will.

## Notes and Further Reading

Neal Curtis, *Idiotism: Capitalism and Privatisation of Life*, London, UK: Pluto Press, 2013.

Arturo Escobar, *Designs for the Pluriverse: Radical Interdependence, Autonomy, and the Making of Worlds*, Durham, NC: Duke University Press Books, 2018.

#113

# Po

Post-Economic

# #113
# A Post-Economic Theory of Design[ing]

*Arturo Escobar*

The fiction of the economy as a domain separate from society and nature is one of the most enduring modern inventions. Design's dependence on this notion has been pervasive and intimate. Decades ago, Raoul Vaneigem (1983), the Situationist, penned the most caustic indictment against it: "*The economy is everywhere that life is not... Economics is the most durable lie of the approximately ten millennia mistakenly accepted as history.*" (Vaneigem, 1994: 17-18). Long before private property showed its ugly head and started to devour territories, people created commons as a strategy to enact their worlds. These worlds, made up of human and nonhuman, material and spiritual beings ineluctably woven together, have continued to persevere, against all odds. "*As the economy's hold weakens, life is more able to clear a path for itself.*" (Vaneigem, 1994: 27-28) It's high time for design[ing] to move decidedly towards a post-economic vision. This would entail a praxis of design as commoning.

## Notes and Further Reading

Raoul Vaneigem, *The Revolution of Everyday Life*, London, UK: Rebel Press, 1983 (2nd Revised edition).

Raoul Vaneigem, *The Movement of the Free Spirit*, NewYork, NY: Zone Books, 1994.

#114

# Of

Of the South(s)

# #114
# An Of the South(s) Theory of Design[ing]

*Alfredo Gutiérrez-Borrero*

The design of the south is customary in regions designed by the hegemonic tradition as 'peripheries'. It finds expression beyond the etymology of the word 'design', where the very idea of etymology is incomprehensible. The design of the south evokes the nouns of its multiple alterities and the verbs of numerous practices. In the worlding practices of that majority of people made up by all minorities. It alludes to the ways in which the people of the polycardinal cultures (from every direction, more appropriate than 'non-western') bring into existence different and deferred artefacts. Among 'exotic', 'barbarian', 'subaltern', 'indigenous' and 'savages' peoples, it pre-figures as design does in Western culture. It brings together into souths (plural) as many design equivalents as there are communities on Earth - geographical, historical, epistemological souths, all of them, like the Southern Pole, represented down under the Globe, places where Nature resists Modernity's invasion and its efforts to separate it from humanity.

## Notes and Further Reading

Ahmed Ansari, Towards a Design Of, From & With the Global South, 2016, available from: https://www.academia.edu/29404507/Towards_a_Design_Of_From_and_With_the_Global_South

Tony Fry, "Design for/by the Global South", *Design Philosophy Papers*, 15, (1), 2017, pp. 3 - 37.

Alfredo Gutiérrez-Borrero, Resurgimientos: sures como diseños y diseños otros. *Nómadas* (Col), (43), pp. 113-129, 2015.

#115

In

Indigenous

# #115 An Indigenous Theory of Design[ing]

*Alfredo Gutiérrez-Borrero*

Old uses engender new materials; new uses engender old materials. Like designing in the Samoan concept of Va - the space between, the betweenness, not an empty space that separates, but a relational space that which brings together entities in a unity-that-is-all, the context that gives meaning to things (Wendt, 1999). To recall territorial bonds by recovering local, that is, indigenous, conditions. To revitalize surfaces with resurgent reliefs. Moving forwards toward journeys that are no longer there, walking backwards toward the future where projects come into existence. To design Mitakuye Oyasin (Lakota - we-all-are-related), where 'we' includes artefacts that extend the scope of our capabilities, improving them with powers inherent to our relationship with fabricated living beings (Vine Deloria Jr., 1990). Designing removing colonial designations in good resurgent life - Minobimaatisiiwin (Anishinaabeg), Suma Qamaña (Aymara), Kume Möngen (Mapuche) etc. Designing material Whakapapas (Maori - genealogies) in a technical course of ancestrality. Mother-stone, father-wood. Giving birth to beings that will give birth to us.

## Notes and Further Reading

Vine Deloria Jr, "Traditional Technology", *Winds of Change*, 5, (2), 1990, pp. 12 - 17.

Albert Wendt, "Afterword: Tatauing the Post-Colonial Body", in *Inside Out: Literature, Cultural Politics, and Identity in the New Pacific*, edited by Hereniko, V., & Wilson, R., 399 - 412. Lanham MD: Rowman & Littlefield, 1999.

#116

# De

Declassified

# #116
# A Declassified Theory of Design[ing]

*Alfredo Gutiérrez-Borrero*

A plural declassified design, intentionally reversible and provisional, requires the de-design of any classificatory order that attempts to fulfil itself as unique and singular. This is because, among humans, sense flows open and unfinished; we see and act more on mutable meanings that we attribute to things than on their physical properties (Krippendorff, 2006: 47), in interdependence. Even similar aspects change for different people and things simultaneously and for the same people and things at different times. Any designed classification imposes transitory orders, susceptible to being altered by a declassification intended to restore creative disorder. For declassified designing, any accepted science is a superstition that denies other superstitions, and each despised superstition a 'science' that encloses the generative potential of denied knowledges. Declassifying then, is not opposed to classifying (that would be "not classifying"); it rather entails reclassifying with other knowledges and purposes (García, 2018: 22).

## Notes and Further Reading

Antonio L. García Gutiérrez. *En Pedazos: El Sentido de la Desclasificación*, Madrid: ACCI, 2018.

Klaus Krippendorff, *The Semantic Turn: A New Foundation for Design*, Boca Raton, FL: CRC/Taylor & Francis, 2006.

#117

# Ge

Gendered

# #117
# A Gendered Theory of Design[ing]

*Elaine Igoe*

Design research academia has proposed a multitude of diagrams in an attempt to model the design process and its associated thinking (Dubberly 2005). Circles, diamonds (Design Council 2005), loops and even 'squiggles' (Newman 2006) have all been used to symbolise this complex process. Modelling the design process and associated thinking in this way, despite the various geometries used to represent it, emphasises linear cognition on a binary, transactional problem-solution pairing. Though if we, as Kimbell suggests (2012: 143), understand design as relational and so generative and matrixial (Ettinger 2006 in Igoe 2013) we move from an epistemology based on a patriarchal, socio-economic context to one which supports a feminist paradigm where design problems and solutions are co-emergent, the subjectivity of the designer(s) and the time, place and context of the design experience is not disregarded and intersectionality can be recognised.

## Notes and Further Reading

Design Council (2005) The 'double diamond' design process model. [online] Available at: <https://www.designcouncil.org.uk/news-opinion/design-process-what-double-diamond>

Hugh Dubberly, ed., (2005) How Do You Design? A Compendium of Design Models. [ebook] San Francisco: Dubberly Design Office. Available at: <http://www.dubberly.com/articles/how-do-you-design.html>

Bracha L. Ettinger, *The Matrixial Borderspace*, Minneapolis, MN: University of Minnesota Press, 2006.

Elaine Igoe, *In Textasis: Matrixial Narratives of Textile Design*, PhD Thesis, London: Royal College of Art, 2013.

Lucy Kimbell, "Rethinking Design Thinking: Part II", *Design and Culture*, 4 (2), 2012, pp. 129 – 148.

Damien Newman, "That Squiggle of The Design Process" [image], 2006, Available at: https://thedesignsquiggle.com/

#118

# Be

Better

# #118
# A Better Theory of Design[ing]

*Craig Bremner & Paul Rodgers*

It might be better for the possible theories of design[ing] if design were to admit to its failure to design a better world. It should be obvious from reading this book that the better world has slipped from our grasp. From the sentimental belief in the design of what-might-become design is now left with what-might-not-become. If the booming design festivals across the world in far-flung locations such as Milan, Graz, Łódź, Gdynia, Bucharest, Prague, Bogota, Lapland, Cape Town, Nairobi, Querétaro, Vancouver, Barcelona, San Francisco, Miami, and many others covering themes such as "Women in Design" and "Polarisation" promising a "better" world through "creativity" are to be believed it would appear there still exists a belief in 'What Design Can Do', but if it can only do what-might-not-become then it must confront the uncomfortable reality that a better way to theorise design is to face up to what design cannot do.

## Notes and Further Reading

Craig Bremner and Paul A. Rodgers, "Design without Discipline", *Design Issues*, Vol. 29, Issue 3, Summer 2013, pp. 4 – 13.

Billy Nolan and Bas Van Lier, *31 Brilliant Ideas for a Better World*, What Design Can Do: Amsterdam, The Netherlands, 2016.

Paul A. Rodgers and Craig Bremner, "The Design of Nothing: A Working Philosophy", In Pieter E. Vermaas and Stéphane Vial (Editors), *Advancements in the Philosophy of Design*, Springer Publishers' Series on "Design Research Foundations", 2018, pp. 549 – 564.

Paul A. Rodgers and Craig Bremner, *Design School: After Boundaries and Disciplines*, Malaga, Spain: Vernon Press, 2019.

Bas Van Lier, Billy Nolan, Femke Van Gemert and Sinette Hesselink, *What Africa can do for Europe - 31 Brilliant Ideas to Inspire the World*, What Design Can Do: Amsterdam, The Netherlands, 2016.

https://www.whatdesigncando.com

# Acknowledgements

Paul and Craig would like to thank all of the amazing contributors to the 118 Theories of Design[ing] book. We have had the privilege of working with all of these contributors in some capacity over recent years, and we are delighted to share their insightful thoughts on theories of design[ing] here.

We are especially grateful to the following people for taking time out of their busy schedules to write one or more theories of design[ing] - Constantin Boym (USA), Maya Dvash (Israel), Chiara Alessi (Italy), Will Holman (USA), Clive Dilnot (USA), Marco Petroni (Italy), Mark Roxburgh (Australia), Saurabh Tewari (India), Jonathan Ventura and Dina Shahar (Israel), Joyce Yee (UK), Mashal Khan (Pakistan), James Fathers (USA), Heather Wiltse (Sweden), Emmanuel Tsekleves (UK), Jen Archer-Martin (New Zealand), Peter Lloyd Jones (UK), Rachel Cooper (UK), Stephanie Carleklev (Sweden), Cathy Treadaway (UK), Delfina Fantini van Ditmar (UK), Danièle Hromek and Jacqueline Gothe (Australia), Nelly Ben Hayoun (UK), Arturo Escobar (Colombia/ USA), Alfredo Gutiérrez-Borrero (Colombia), and Elaine Igoe (UK).

We would also like to thank our fantastic colleagues, past and present, at institutions including Strathclyde University, Imagination@Lancaster University, Northumbria University School of Design, Edinburgh Napier University, Cambridge University, and Charles Sturt University.

In particular, we would like to acknowledge the support and generosity of many talented individuals that we have worked with and met on our design research journeys. Here, special mention is due to Professor Rachel Cooper, Dr Giovanni Innella, Andy Tennant, Professor Ranulph Glanville, Adjunct Professor Martin James, and Adjunct Associate Professor Ian Coxon.

Last, and certainly not least, Craig would like to thank Laura and Elio, and Paul would like to thank Alison, Charlie, and Max for their never-ending support.

# Contributor Biographies

**Paul A. Rodgers** is Professor of Design at the University of Strathclyde, Glasgow. He is also the current Arts and Humanities Research Council Leadership Fellow for Design in the UK. Before this, he was Professor of Design at Imagination@Lancaster University, Professor of Design Issues at Northumbria University, School of Design, Reader in Design at Edinburgh Napier University, and a Research Fellow at the University of Cambridge's Engineering Design Centre. He has had an extensive career in design research with over 20 years of experience in product design research and development. He has led several research projects for Research Councils in the UK and design projects funded by the Scottish Government and The Lighthouse (Scotland's National Centre for Architecture, Design and the City). He is the author of more than 150 papers and eight books, including The Routledge Companion to Design Research (2014). His current research explores the discipline of design and how disruptive design interventions can enact positive transformational change in health and social care and elsewhere.

**Craig Bremner** is Professor of Design at Charles Sturt University, Australia. Prior to this, he was Professor in Design Pedagogy at Northumbria University, UK, and before that Professor of Design at the University of Canberra, where he was also Dean of the Faculty of Design and Architecture. He holds a BA in Literature (UWA), a Masters in Design (Domus Academy, Milan) and a PhD (RMIT). His research deals with developing methods to discover how and why we don't know much about the experience of design, as well as finding ways to clarify the reason why 'not-knowing' is an essential and valuable beginning point of practice. Some applications of his research methods have traced the experience of living in Glasgow, using banks and driving motorcars. In his private practice, he has curated design exhibitions in Australia, the USA, and Japan, and he has worked as a designer in Italy, Scotland and Australia.

**Chiara Alessi** is a journalist and curator in the field of design. She writes for many of the major magazines in this area, such as Domus, Interni and Klat. For several years she has been studying the new Italian design culture and its impacts and implications. On this subject, she lectures in some of the most important schools and universities. Recently she has published

"After the '00s. The New Italian Design" (Editori Laterza, 2014) and "Design without Designers" (2016), an essay and survey about the 'other' jobs in the Italian design sector. The independent video documentary "Travelling Commonplaces and Special Trades" is dedicated to this subject.

**Jen Archer-Martin** is a spatial designer, thinker and educator. Her transdisciplinary practise encompasses design, writing and performance, exploring spatial and material ecologies of care. Recent work includes *taking note(s)_performing care,* and collaborations *Make/Use* and *bit-u men-at-work*. Jen is a lecturer and coordinator of Spatial Design at Massey University, Wellington, New Zealand. She is of English, Scottish and Māori (Ngā Puhi) descent.

**Constantin Boym** is Head of Industrial Design at the Pratt Institute in New York City City, USA. Constantin Boym was born in Moscow, Russia in 1955, where he graduated from the Moscow Architectural Institute. In 1984-85 he earned a degree of Master in Design from Domus Academy in Milan. In 1986, he founded Boym Partners Inc. in New York City City, which he runs together with Laurene Leon Boym. Boym Partners Inc. brings a critical, experimental approach to a range of products and environments that infuse humour and wit into the everyday. The studio's designs are included in the permanent collection of the Museum of Modern Art in New York City. In 2014, Constantin Boym received an honorary doctorate from the Corcoran College of Art and design. From 1987 to 2000 Boym was a teacher and program coordinator at Parsons School of Design. In 2010-12 Boym served as Director of Graduate Design Studies at Virginia Commonwealth University in Qatar.

**Stephanie Carleklev** is a graphic designer and senior lecturer at Linnaeus University in Sweden. Her work is driven by a strong interest in life and what sustainability could mean. As a course and former programme leader, Stephanie primarily exercises her design skills by designing education. Her most recent work and research focus on time and its potential in design for sustainable change.

**Rachel Cooper OBE** is Professor of Design Management at the University of Lancaster, where she is Chair of Lancaster Institute for the Contemporary

Arts and also Imagination@Lancaster (a centre for research into products, places and systems for the future). Her research interests cover design management, design policy, new product development, design in the built environment, design against crime, and socially responsible design. Between 2003 and 2008 she led 'Vivacity 2020: Sustainable Urban Design for the 24-Hour City', a £3m EPSRC funded project over five years looking at Manchester, London, and Sheffield. She has authored several books including "The Design Agenda" (1995), "The Design Experience" (2003), "Designing Sustainable Cities" (2009), "Constructing Futures" (2010), "The Handbook of Design management" (2011) and is currently the commissioning editor for an Ashgate series on Socially Responsible Design. Professor Cooper is President of the European Academy of Design, and Editor of The Design Journal. She was a member of the UK Research Assessment Exercise Panel for Art and design in 2008, and in 2011 she was invited to be a member of the European Design Innovation Initiative Leadership Board that produces recommendations on Design for Prosperity and Growth for the EU.

**Clive Dilnot** is Professor of Design Studies at the Parsons School of Design and The New School in New York City. Clive Dilnot was educated as a fine artist, and later in social philosophy. He has taught worldwide including at Harvard University, the School of the Art Institute in Chicago and in Hong Kong, as well as in Australia and the UK. Publications include Ethics? Design? (Archeworks, 2005) the essay for Chris Killip's Pirelli Work (Steidl, 2006) and the co-authored Design and The Question of History (2015). He is the editor of A John Heskett Reader: Design History Economics (2016) and of Heskett's seminar on design and economic Thought, Design and the Creation of Value (2017). He is currently working on a four-volume series Thinking Design: History; Ethics; Knowledge; Configuration (2019-20). He is founding editor of Designing for Dark Times/The Urgency of the Possible, a new series of short books and polemical essays, and Radical Design Thinkers, re-publishing significant texts in design thinking since 1960.

**Maya Dvash** is Chief Curator at the Design Museum Holon in Israel. Throughout her time at Design Museum Holon, Dvash has curated a vast array of design exhibitions and written on design for numerous platforms. In addition to her functions as editor, curator and writer, Dvash lectures at leading design academies in Israel. Prior to her career at Design Museum Holon, Dvash held various leading editorial positions in some of the top-tier publishing houses in Israel (Kinneret, Zomora-Bitan and Modan). Dvash was then appointed Chief Editor of Binyan v'Diyur (Building & Housing) magazine, a seat she filled for six years. Dvash holds a Bachelor's degree in Literature and a Master's degree in Art and Curatorship from Ben Gurion University.

**Arturo Escobar** is Kenan Distinguished Professor of Anthropology at the University of North Carolina, Chapel Hill, USA. His academic research interests include political ecology, anthropology of development, social movements, anti-globalization movements, and postdevelopment theory. He holds a BSc in chemical engineering from the University of Valle, Cali, Colombia, and he completed one year of studies in a biochemistry graduate program at the Universidad del Valle Medical School. He has a master's degree in food science and international nutrition from Cornell University and an interdisciplinary Ph.D from the University of California, Berkeley, in Development Philosophy, Policy and Planning. He has taught mainly at universities in the USA, including the University of Massachusetts Amherst, but also abroad at institutions in Colombia, Finland, Spain, and England.

**Delfina Fantini van Ditmar** is a design researcher and lecturer whose work investigates the socio-ecological and political implications of technological 'smartness' when applied to human bodies, homes and cities. Delfina has a BA in Biology and completed a year of an MFA at Konstfack University in Stockholm. She holds a PhD from the RCA with a thesis entitled 'The IdIoT', which investigated the socio-political implications of technological 'smartness' and the algorithmic processes, characterised as the 'Algorithmic Paradigm'. Delfina has a transdisciplinary background linking design research, critical algorithmic studies, architecture, biology and the sociology of technology. She has been a visiting lecturer in several institutions, including The Bartlett, Architectural Association, Canterbury University, Liverpool University and TU Berlin, among others.

**James Fathers** is Director of Syracuse University School of Design and the Iris Magidson Endowed Chair of Design Leadership. Fathers' teaching and learning activity focuses on sustainability, universal design, and design in a development context. His research interests lie in socially responsible design; his doctoral research focused on the role of design in a development context, which led to a 12-month research sabbatical to India, where he worked alongside local crafts groups to develop appropriate design training strategies to facilitate enterprise development.

**Jacqueline Gothe** is a design researcher in visual communication design and the Director of Visual Communication Design in the School of Design at University of Technology Sydney, Australia. Her research approach emphasizes research through design as a knowledge creating paradigm. Jacqueline has widely researched the application of communication and design principles in the natural resource management sector, investigating transdisciplinary approaches in projects dealing with the consequences of environmental flows and pesticide toxicity on the Hawkesbury Nepean River. Her PhD, awarded in 2016, investigated the role of the visual communication designer in complex interdisciplinary and cross-cultural environmental communication design projects. Since 2011 she has worked in partnership with Firesticks, an Indigenous-led network that mentors, shares and supports the revival of cultural burning practices in natural resource and fire management contexts. Her communication and information design contribution to Firesticks has been recognised in 2011 and 2014 by the International Institute of Information Design Awards (IIID).

**Alfredo Gutiérrez-Borrero** is Professor of Industrial Design at the Universidad Jorge Tadeo Lozano in Bogotá, Colombia. He is interested in opening 'the future' to other pasts and 'the past' to other futures. Gutiérrez explores the confluence between Western design and archaeology (the only ones, to tell the truth) and the polycardinal prefigurative forms (from outside the West) that he calls dessobons.

**Nelly Ben Hayoun** is a French designer, an award-winning director and experience designer; she works with scientists and engineers to devise events and experiences. Nelly is also an exhibitor and keynote speaker who has worked with museums and design centres across the world. In 2013, Icon Magazine nominated Nelly as one of the 50 international designers "shaping the future". In 2014, Wired Magazine awarded her with a

WIRED Innovation fellowship for her work to date and for its potential to make a "significant impact on the world". In 2015, Nelly was nominated for a Women of the Year Achievement Award. Also, in 2015, she released her feature film Disaster Playground. The film is based on an investigation of emergency procedures for disasters such as earth-bound rogue asteroids. In 2016, she began work on her next project: feature film, digital platform and exhibition entitled "The Life, the Sea and the Space Viking". Nelly is a member of the International Astronautical Federation, Space Outreach and Education committee.

**Will Holman** was educated as an architect at Virginia Tech and the Rural Studio. Will is the Executive Director of Open Works, a makerspace that opened in central Baltimore in 2016. He is the co-founder of the Industrial Arts Collective; their mission is to be an online resource for communication, collaboration, and general education on all the amazing people living and working in Baltimore. Will is author of Guerilla Furniture Design, which is an innovative guide to dozens of strategies for upcycling scrap cardboard, metal, plastic, or wood into dependable shelving units, sturdy tables, and fun lamps. Most recently he has contributed two essays: The Toaster Paradox and The Open Source Object to the Open Making Manifesto: Field Guide in conjunction with the Open Desk exhibit at the Vitra Design Museum.

**Danièle Hromek** is a spatial designer, speculative designer and public artist, fusing design elements with installations, sculptural form and research. Danièle also works as a researcher, educator and cultural advisor considering how to Indigenise the built environment by creating spaces to substantially affect Indigenous rights and culture within an institution. She works at the intersection of architecture, interiors, urban design, performance design and fine arts. As an Aboriginal researcher and designer, her work is grounded in her cultural and experiential heritage, often considering the urban Aboriginal condition, the Indigenous experience of Country and contemporary Indigenous identities. Gaining her experience globally, she has lived in London, Paris, Barcelona, Vancouver and Sydney. Danièle is a Saltwater woman of the Budawang tribe of the Yuin nation, with French and Czech heritage. Danièle trades as Djinjama, a word in Dhurga that means 'make or complete or produce or build something'.

**Elaine Igoe** is a Senior Lecturer in Textiles and Fashion at the University of Portsmouth as well as a visiting tutor at the Royal College of Art. She has acted as an external adviser for doctoral students at the University of Portsmouth, the Royal College of Art and Chelsea College of Art and Design. She holds a Fellowship of the Higher Education Academy and is a full member of The Textile Society and the Design Research Society. Her research, both practice-based and theoretical, has always taken an approach that is experimental and conceptual; questioning and exploring the innate and typical characteristics of textiles, materials and cloth, particularly in the context of the body. Her theoretical work in design research utilises feminist qualitative research methods, exploring relationality through storytelling, autoethnography and metaphor.

**Mashal Khan** is a strategic designer and visual researcher currently working for Kaarvan Craft Foundation, an NGO based in Pakistan, striving to empower women in low-income communities through life skills. Mashal sees the world through a sense of wonder, humility and respect – ever ready to confront complex social issues by designing new ideas, tools, methods and action agendas.

**Peter Lloyd Jones** completed his PhD in cellular and genetic pathology at Cambridge University, followed by post-doctoral fellowships at UC Berkeley and the University of Toronto. In 2005, Peter became a tenured Associate Professor of Pathology and Lecturer in Architecture at The University of Pennsylvania. Peter Lloyd Jones is an award-winning cell and molecular biologist and inventor, whose discoveries have uncovered fundamental mechanisms in embryogenesis and human disease, including breast cancer, lung development and pulmonary hypertension. Peter's work constantly seeks and uncovers new solutions to complex problems in biology, medicine and design via extreme collaborations with diverse individuals from multiple fields, including mathematics, computation, chemical engineering, and industrial, fashion, textile and architectural design. Peter is a Professor of Design at Sheffield Hallam University, UK. Before this he was the first Associate Dean of Emergent Design and Creative Technologies at The Sidney Kimmel Medical College at Thomas Jefferson University (TJU), where in 2013, he became Executive Director of MEDstudio @JEFF; an anti-disciplinary research and education unit which is the first of its type within a US medical school, and is focused on discovering new and dignified solutions in healthcare using approaches deeply rooted in empathy.

**Marco Petroni** is a design theorist and critic. He is also Adjunct Professor at Politecnico (Milan) and the Università della Campania (Naples). He is Curator at large at Plart Foundation (Naples). He collaborates with several art, architecture and design magazines such as Domus, FlashArt, and Artribune. Petroni studied contemporary art and architecture. He develops innovative curatorial projects and events related to the design world themes with a transdisciplinary approach.

**Mark Roxburgh** is Associate Professor of Design at the University of Newcastle, a management stooge, and a fading would be indie rock star (google Joeys Coop). Mark's scholarly interests cover design research, visual communication theory and practice, and photographic theory and practice. His PhD explored the central role that visual images and visual perception play in design, with a specific emphasis on how photographic images condition us to perceive, experience and transform the world in a self-replicating manner. His more recent pursuits have been: developing a phenomenological theory of photography to counter the dominance of critical theory and semiotic deconstruction; developing a theory of design as a form of embodied perceptual synthesis to counter the dominance of the design problem-solving metaphor; and making music that generally gets ignored to counter the dominance of youth-driven culture. More recently he has grown suspicious of the instrumentalization of the anthrochauvinist bandwagon known as human-centred design and thinks it time to head to the hills and dig the bunker.

**Dina Shahar**, a graduate of the Royal College of Art, London (MA RCA), is Head of the Inclusive Design Department, at the Hadassah Academic College, Jerusalem. Her academic interests span the methodologies of Inclusive Design and the evolving roles of the design discipline as a whole. Her creative work focuses on design for public spaces.

**Saurabh Tewari** is Assistant Professor at the School of Planning and Architecture Bhopal. He studied Architecture (B. Arch, 2003-08) at SSAA

Gurgaon and Communication Design (M. Des, 2008-10) at IDC IIT Bombay. He has been teaching basic design, graphics, and architectural history for eight years. He is a PhD candidate at Design Programme, IIT Kanpur researching on design and its history in India.

**Cathy Treadaway** is Professor of Creative Practice at Cardiff Metropolitan University and a founder member of the Centre for Applied Research in Inclusive Arts and Design (CARIAD). She is Principal Investigator on the AHRC 'LAUGH' design for dementia research project. Cathy is a Fellow of the Royal Society of Arts and a Fellow of the UK Higher Education Academy.

**Emmanuel Tsekleves** leads research at the intersection of design, health, wellbeing and technology at Imagination@Lancaster at Lancaster University. He conducts design research by working with communities to develop new ways about health, wellbeing and technology to create a 'culture for health' that is knitted into everyday community life. He is the co-editor of the Design for Health book published by Routledge.

**Jonathan Ventura** is a design anthropologist, specializing in design theory, research and practice in the fields of social and healthcare design. Jonathan is a senior lecturer at the Department of Inclusive Design at Hadassah Academic College and the Graduate Program of Design at Shenkar - Engineering, Design, Art and a visiting researcher at the Royal College of Art's Helen Hamlyn Centre for Design.

**Heather Wiltse** is Assistant Professor at Umeå Institute of Design, Umeå University, Sweden, where she is also currently serving as director of PhD studies. Her transdisciplinary research centres around trying to understand and critique the role of digital, networked, computational things in experience and society in ways that can inform design.

**Joyce Yee** is Professor of Design and Social Innovation at Northumbria University Design School, UK. She holds a BFA in Graphic Design, an MA in Visual Communication and a PhD in Design. Her research focuses on the role, value and impact of design in areas of service design and social innovation. She is also interested in the role of practice as a form of inquiry and the inherent 'messiness' of design methods.

# Index

## C

## D



## E

## F



## G

## N



## O



## P

## Q



## R

## S

## Y



## Z



## #

www.ingramcontent.com/pod-product-compliance
Lightning Source LLC
LaVergne TN
LVHW010722210926
848485LV00007B/1417/J

* 9 7 8 1 6 2 2 7 3 9 6 2 2 *